Studies in International Relations

Edited by
Charles MacDonald
Florida International University

A Routledge Series

Studies in International Realtions

Charles MacDonald, *General Editor*

Promoting Women's Rights
The Politics of Gender in the European Union
Chrystalla A. Ellina

Turkish Foreign Policy and Turkish Identity
A Constructivist Approach
Yücel Bozdaðlioðlu

Organizing the World
The United States and Regional Cooperation in Asia and Europe
Galia Press-Barnathan

Human Rights in Cuba, El Salvador and Nicaragua
A Sociological Perspective on Human Rights Abuse
Mayra Gómez

Negotiating the Arctic
The Construction of an International Region
E.C.H. Keskitalo

The Common Fisheries Policy in the European Union
A Study in Integrative and Distributive Bargaining
Eugénia da Conceição-Heldt

Malaysia and the Development Process

Globalization, Knowledge Transfers and Postcolonial Dilemmas

Vanessa C. M. Chio

Routledge
Taylor & Francis Group
LONDON AND NEW YORK

Published in 2005 by
Routledge

2 Park Square, Milton Park, Abingdon, Oxfordshire OX14 4RN

711 Third Avenue, New York, NY 10017

Routledge is an imprint of the Taylor and Francis Group.

Transferred to Digital Printing 2005

First issued in paperback 2012

Library of Congress Cataloging-In-Publication Data
Chio, Vanessa C. M., 1963–
Malaysia and the development process : globalization, knowledge transfers and post-colonial dilemmas / Vanessa C. M. Chio.
p. cm.—(Studies in international relations)
Includes bibliographical references and index.
ISBN 0-415-94941-6 (hardback : alk. paper)
1. Malaysia—Economic conditions. 2. Malaysia—Social conditions. I. Title. II. Series: Studies in international relations (Routledge (Firm)).

HC445.5.C48 2004
330.9595--dc22 2004009235

ISBN 13: 978-0-415-64985-8 (pbk)
ISBN 13: 978-0-415-94941-5 (hbk)

To mindfulness and grace,
to the memory of a father,
and the loving support of a family.

Contents

List of Figures

List of Terms

APO	Asian Productivity Organization
FDI	Foreign Direct Investment
HRDC	Human Resource Development Council
HRDF	Human Resource Development Fund
ILO	International Labor Organization
IMF	International Monetary Fund
IMP	Industrial Master Plan
INTAN	National Institute of Public Administration
ISO	International Standards Organization
ITAF	Industrial Technology Assistance Fund
MAMPU	Malaysian Administrative Modernization and Planning Unit
MCA	Malayan Chinese Association
MIC	Malayan Indian Congress
MITI	Ministry of International Trade and Industry
NDP	New Development Policy
NEP	New Economic Policy
NPC	National Productivity Corporation
QCC	Quality Control Circle
SAP	Structural Adjustment Policy
SIRIM	Standards and Industrial Research Institute of Malaysia
SMI	Small and Mid-Size Industry
TQM	Total Quality Management
UMNO	United Malays National Organization
UNDP	United Nations Development Program

Foreword

How to know "the other"? This is, arguably, one of the most fundamental questions in contemporary social science scholarship. Research is, after all, always already about/for "the other" and, therefore, the possibility of knowledge acquired through research is unavoidably involved in all the intellectual and analytical imbroglios which "otherness" implies. More pointedly, however, "the other" is the central signifier for all knowledge that claims to be "international" (or, as of late, fashionably "global"). Without "the other" there is indeed no international knowledge to be had.

Reconsidering the latter statement is the main task of this book. Conceptually, methodologically and rhetorically, Vanessa Chio guides the reader through the maze of difficulties implied in understanding "otherness" in international management contexts. The book provides both a problematization of assumptions that are often taken for granted in international inquiry, as well as alternative approaches through which it is possible to continue these inquiries while evading some of their most serious pitfalls. And it does so while interrogating, once and again, the illusory conditions which make us believe that, at the end, "we will know the other."

The book is also about understanding contemporary Malaysia as it negotiates its position in the global economy. In a journey into diverse scenarios constituting the country's "development path" the author takes the reader on a tour filled with local color, vividly telling stories which range from a taxi ride through crowded streets to management training programs for factory personnel; and from personal memories of the author's childhood to the "official story" as told by central government dignitaries. As the journey unfolds the story becomes more complicated. The multiplicity of narratives illustrates the intricate space about which there is no seamless story to tell. Rather, culture and history contain the local and the global, reflections and refraction, the same and the other. From this perspective, knowing Malaysia as "the other" becomes a more arduous enterprise, for at every possible moment in which one might ask, Who are they? It is also possible that one might be asking, Who are we?

This, of course, is precisely what the author intends to convey. We live at a moment in time in which the West increasingly focuses on "the rest" as a last frontier for economic development. This focus is neither benign nor particularly well informed. The literature on development as well as on international management, mostly of Anglo-American vintage, has produced over the course of the years images of "the rest" as "exotic and primitive," "childlike," "devoid of agency," "submissive and nimble" and so on and so forth. These images, often reflective of a colonial past either real or imagined, have maintained their currency in spaces as disparate as Hollywood movies and cross-cultural management scholarship, the contemporary heirs of travel chronicles and literary and historical productions of more than a century past. The "others" so represented are always distant, behind the West in knowledge, empty vessels waiting to be filled. They are, more often than not, silent and invisible, acquiring visibility only if animated by the "winds of progress" blowing in their direction, their voices articulated only "in translation."

Throughout its various chapters, this book produces a series of counter-representations to those above. The "others" are neither "behind" nor "empty of meaning"; they are neither silent nor invisible. Their lives are full of active engagements with a world as current as that of any other place today; they learn what is to be learned but in the process of learning they are also actively constructing other possibilities and meanings, changing, innovating and constructing other knowledge. Power relations permeate these narratives. The actors negotiate globalization processes at the most local of everyday life. They accept and reject, submit and resist. Altogether these counter-representations offer a window into the lived experience of the global economy for those whose life is at the interface of such processes. They are the inhabitants of "contact zones" created by globalization, neither here nor there and soon to be extended to the rest of our world. From them we learn.

And yet, at another level, this book is a meditation on approaches at the disposal of those who, no matter from where they hail, are engaged in conducting international scholarship in the model of the West. In particular, the book calls attention to the extensive interdisciplinary repertoire available to today's researchers who, if so inclined, can address their subject matter well beyond the universalizing and colonialist implications of social sciences theories passing themselves off as neutral. To be admired, indeed, is the depth of scholarship represented by this book. The reflexivity brought about by post-structuralism, where "otherness" came into play as the necessary condition for the articulation of knowledge, has opened the door to forms of interrogation in postcolonial studies through which "the other" has become, in fact, no longer silent or invisible. These theoretical approaches are central inspirations in the configuration of the book's argument. Chio weaves her stories mobilizing theoretical positions from the humanities as easily as she does from the social sciences. Postcolonial analysis, institutional ethnographies, anthropology of development as well as international management and organization studies, compose a theoretical and methodological mosaic which illustrates many possibilities open to contemporary researchers and offer, at the same time, guidance on how to go about it.

In sum, in our view this book is at the vanguard of contemporary organizational and management scholarship addressing international contexts and globalization processes. It shows the limitations of traditional managerial scholarship for understanding a world that can no longer be explained through the teleology of modernity, but it also offers ways to overcome such limitations. The author is, as well, a good representative of a new breed of organizational scholars, who through a critical awareness of scholarly developments beyond the narrowness of their fields have been able to bring together new theorizations and methodological possibilities that make our field, once again, an exciting place from where to engage with the world and, perhaps, try to make it better for the many.

Marta B. Calás and Linda Smircich
University of Massachusetts at Amherst

Acknowledgments

There are moments when the blessings that heretofore are overlooked, taken for granted, hidden or unknown and unrecognizable seemingly emerge in all their guises and forms and remind us about all that has been given, and all that has been received. This is one such moment. As I sit contemplating the particular road that led me to this point where I am writing the acknowledgements to a story long in the making, I am reminded of the journey that is a life, and the many journeys that we take in life. Along the way, I have learned and relearned the concept of community as I lived instead, the meaning of community.

Like the mustard seed that contains a universe in it, this book is an agglomeration of multiple places and multiple faces. From the shores of Malaysia to the institutional space that I currently reside in, this text and this story contain the traces and the presence of people and events, history and memory, encouragement and support that make it and I what we both are at this moment. It would not be in the absence of all that has gone into the constitution of this self and this story. To these people and events, this history and memory, this encouragement and support: my most joyful and deep-felt appreciation for what has been given, and what has been received.

Countless are those in Malaysia for whom this thanks must be conveyed: the managers and officials, staff members and workers, colleagues and friends. This book is a humble testament to their help and support. While most must remain anonymous, I would like to acknowledge the warm welcome given me by Beda Lim, Jennifer and Peter Ho, Poh Hing and Jennifer Lee. Traversing halfway across the world, I thank Multitech Inc. [name changed to preserve anonymity] for its permission to reprint the corporate orientation video I quote later in this text.

Equally pertinent to the existence and success of this endeavor is the support, the teachings and the suggestions of Arturo Escobar and R. Radhakrishnan. Not only are these much appreciated, their writings have been invaluable for the insights and understanding that have gone into the constitution of this work. I thank them for the stories they both tell, and tell so well, in their respective disciplines.

To Marta Calás and Linda Smircich, two who are pillars of my community, I thank them for their unstinting and manifold support throughout the years; support that is way above and beyond what is required in their respective locations as teachers and committee members. This has been essential for my ability to tell the story I tell here. Not only do their writings provide an entry-point for the investigation I undertake in this book, their hard work, commitment to scholarship and faith in my capacity serve as constant reminders of what is and what can be. Marta—a special word of thanks here for the time and effort put into this manuscript through its many stages and incarnations. Most of all, I thank them for their generosity in including me into their lives and home. This will not be forgotten anytime soon. No matter where I am, or what else I may be, my time amidst the towering pines in front of the house, the greenery and color in the yards, and sharing of lives and meaning will forever remain a part of what and who I am. And to Jacques D. himself, I salute you as always!

I thank here Anthony Butterfield and Larry Zacharias for their support and cheer throughout my time in the field and beyond. Also to be acknowledged are those friends who make up my global extended family: friendships and relations consolidated by the vagaries of graduate school, common interest in scholarly questions, and the rarefied—and extremely bountiful—experience that is food! To Bobby Banerjee, Kuldhir Bhati, Patricia Clarke, Massimo Fussilio, Carlos Gonzalez, Roberto Henke, Jacqueline Hodgson-Blackburn, Karen Landers, Wayne Millette, Raza and Ali Mir, Mike Morris, Louise Peel, Diana Wong, Lisa Robinson, Usha Rao, Mario Zappacosta and like others not mentioned here—may the bounties unbound yet more.

To my friends and colleagues, and staff members at the University of Washington, Tacoma—both in and outside the Milgard School of Business—my grateful appreciation for the welcome and support that has been extended to me. To series editor Charles MacDonald and my Routledge editor, Kimberly Guinta, many thanks indeed for your support and suggestions on this manuscript.

Finally, in ending this acknowledgement, I thank my family for their unquestioning support and faith throughout all these years. I cannot begin to describe the gift that is the freedom they have given me to wander the world, secure as ever in the amplitude of their care, even though I know there were times when they wondered over the whereabouts and whatnots of this peripatetic nomad. Beyond this simple message, I leave unvoiced but ever aware of the magnitude of the gift that is they.

Chapter One

Introduction

Speeding along the multi-lane highway connecting Subang International Airport to the heart of Kuala Lumpur, Malaysia's capital and soon-to-be home to the world's tallest building, there is a certain sense of having arrived. More specifically, as you encounter the landscape and activity of fast-expanding and fast-changing KL [as the capital is known to locals], you notice and feel the tremendous amount of hustle and bustle permeating its every corner, an almost non-stop conglomeration of motion and noise, and you think to yourself: "things are happening here. . . ." The symbol of Malaysia's entry into the global economic arena, the capital is a city constantly in motion. Seemingly unending streams of cars, motorbikes and pedestrians vie with buildings under constant and continuous construction for preeminence within the newly arrived person's sensory perception—antlike platoons of beings going somewhere, anywhere and everywhere.

Somewhat like the busiest of cities elsewhere in Asia, the degree of movement and noise is at once invigorating and annoying in its constancy and frequency: the blaring of horns as drivers cut and overtake each other, the rumbling of pink city buses filled to bursting with passengers, the sound of steel girders being hammered into the earth, the constant stopping and going of traffic making way for pedestrians and road-hogs alike. And yes, even the occasional yell or two [to put it in its secular form] from these denizens can be heard through the insulated confines of air-conditioned taxis and cars. However, unlike stereotypical images of Asian cities of old, gleaming steel towers dotted and concentrated throughout this landscape glare and wink down at passersby, with yet more of these like structures under construction still.

Meanwhile, palm fronds and the lushness of vegetation flanking the Airport's entrance to the highway and the humidity that rushes out to meet the newly arrived signal a certain geographical displacement. Selamat Datang! Welcome! This, along with the somewhat disorienting effect of seeing traffic organized along the "wrong" side of the road reinforces a sense of spatial displacement, just as women in colorfully printed baju kurungs and kebarungs, road signs or names in Malay, and glimpses of songkoks and turbans signal a certain cultural displacement. The visitor realizes that he/she is not home, the returnee a slow but growing sense of orientation as the body and senses register sights, sounds and smells vaguely familiar yet not quite so.

Once on the road, however, as you speed by the Sungai Way Free Trade Zone located about halfway between the Airport and the city center, the mind's eye register names familiar to visitors the world over: Toshiba, Motorola, Matsushita, Panasonic. And so on the list goes—flagships of transnational and global capital denoting modern technology and goods; propellers of the nation's export industrialization strategy; symbols of development and progress in this, the New World Era. Not so different after all.

A second glance and you feel—physically and emotionally—the "beat" behind the motion: a certain alertness and determination, a sense of positive-ness and the boundless possibility attached to possibility itself—reactions intoxicating not just for their capacity to uplift the spirit, but also for their commentary about the worlds and homes left [temporarily] behind, where images of other gleaming steel towers vie with images of the homeless and the unemployed, of well-insulated museums and drug-infested ghettos.

Yet another glance and you notice the dustiness and fogginess of emissions polluting the skies, the redness of earth exposed, and the impatience amidst the optimism. You notice vaguely-remembered buildings and streets missing; suburbs no longer isolated but linked to other like suburbs, most barely distinguishable from one another to the unfamiliar eye: all interrupted now and then by still more buildings and steel structures. You "see" the absence of cows and corner newsstands; the presence of a highway cutting through what used to be the grounds of the Istana, or rather, the National Palace.

* * *

Welcome to a panoramic view of Malaysia in the act of striding towards an Industrialized[1] status; the reason for my presence in this capital in this instance. Indeed, the return at this juncture to my nation of birth—albeit on a somewhat different level [researcher], with a somewhat different purpose [research], from a somewhat different location [western researcher, management scholar]: definitely not the usual jaunt home for a visit with family and friends. This then is the terrain and site from wherein I will be living, interacting and researching for the year to come. It is January 1995. Unbeknownst to me, this simultaneous sensation of familiarity and disorientation was to become an almost constant part of my research time, location and process during the coming year. While glimpses of these shifts had been seen during my previous visits to negotiate the possibility of returning, they were to become central to the project itself: an index of the very processes I myself was trying to document at a more institutional and sociological level.

Leaving the country at a young age for the West, the Malaysia I inhabited during this time was one which was somewhat alien to my experience as a Malaysian up to this point—filtered as it was with the institutional and economic exigencies of a nation and a society in the midst of, literally and figuratively, transforming itself. Yet this very institutional and economic exigency was a key reason for my return. It is one of the ironies of the project that the information and tools by which I was equipped to assess this transformation were arsenal gleaned from my time in the West: whether as a participant, scholar and/or critic of the organizational and social sciences. So it was that my return would be marked in the most immediate and fundamental of ways by this paradoxical juxtaposition of realities; filtered throughout by the memory of a Malaysia from my childhood, and a simultaneous exposure to what it means to live on an everyday basis in this nation as a researcher in the community: a personal and social Other in opposition to the "official" Malaysia I had come to try and understand.

From my starting point within two Western electronics multinationals in the states of Selangor and Penang to the varied numbers and levels of statutory, educational and government agencies which I traced through in my documentation of the epistemic and institutional forms and practices propping-up the nation's development strategy, the world I traversed was the world of technocratic expertise. It was a world peopled by the presence of senior and high-level engineers, administrators, scholars and managers; where overriding concerns dealt with overseeing production runs, meeting policy objectives, and translating organizational aims. Whether private or public, local or global, all the individuals I met and talked with during this time were concerned in one way or another with the formation and sustenance of institutions.

Yet, as I discovered time and again throughout my stint in Malaysia, even within the otherwise technocratic and linear spaces of the foreign multinationals, there were certain moments infused with local mediations, appropriations and contestations over the so-called dominant rationalities and practices of modern institutions. As a commentary on culture and volition, these mediations and appropriations are themselves indicators of the plurality and agency of a people and a society already cultured, historicized and filled with meanings.[2]

I highlight this technocratic focus here to acknowledge the class-based and professional character of the world that the project and I are located in. I acknowledge this to claim a certain ownership over the subject and location of my inquiry, and begin making visible the academic and conceptual boundaries organizing the subject itself. Included here are its focus on Economics as a terrain of production and way of knowing, its overriding interest in productivity, and the explicit privileging of administrative efficiency.

Mine was not the world of the operator within these multinationals. Nor was it the world of the foreign worker negotiating these same multinationals. As I realized soon after beginning my fieldwork, I could no more negotiate—or be accepted *by* and/or be accepted *as* representing—this other world, than I could climb the Himalayas. The reason? I am not one of them. And while I was sure that I could—technically, administratively, practically—take on the job of an operator, I could never paradoxically, *be* him/her. History and a whole slew of other demarcations (e.g., class, education, ethnicity, culture) became a boundary locking me out, and them in.

Existing in tandem with this official and located visibility was a world colored by the constant battle to outwit traffic jams and taxi drivers; of finding myself ignominiously and repeatedly lost amidst the concrete cookie-cutter jungles of suburban Subang Jaya. *Aduh miss, if you don't know where you live, how I take you home?* To which this researcher quickly and apologetically replies: *excuse, excuse, sorry encik, I baru balik from overseas. Yaakah? okay, okay, tak apalah, I cari your home, don't worry.* It is a refrain and an excuse that, conveniently and inconveniently for me, worked and lasted for six months across two states. The assumption and existence of an all-knowing, purposeful researcher-anthropologist reduced to the rubble of quick apologies and kindness from strangers; the authority of the researcher-expert replaced by spatial and social disorientation.

This social world of the researcher I was also plagued by an endless negotiation between seemingly inchoate identities [Western researcher, Malaysian-born Chinese], mis-identities [non-Chinese, non-local Chinese, non-local Malay, non-Malaysian], or much to the amusement of locals and friends alike, of having to deal with good-natured ribbings about my *Mat Salleh* Malay, my "twangy" Cantonese, and my hopeless Hokkien. It was a world colored by constant reports of landslides and construction accidents, discussions about religion, and questions about the Malay language and Malaysian identity. I also witnessed from up-close friends and colleagues juggling between home and work, earning power and escalating educational and medical costs, and foreign maids negotiating from the private confines of Malaysia's growing suburbia. Most of all, it was a time marked by a concerted effort on my part to learn and relearn the living history of the nation.

The challenge for me became one of trying to understand from the ground up and from the standpoint of lived experience—the agglomeration of knowledge/knowing and location that is "I/eye"—how these two or more worlds, these "malaysias" were connected. I found myself asking time and again: how are decisions made in institutions, the concepts or practices used by decision makers, and the links forged between institutions and sectors related to the congestion, the perceived absence of cows, the boundless optimism, the search for identity, and the pride and the frustration that I was myself witnessing and/or experiencing during this same time? Just as the challenge was simultaneously one of trying to understand from this standpoint of a living history how these "malaysias" were/are connected [economically, socially, academically, theoretically] to the West I too had come from and am traversing still.

But this journey through paradox and irony, theory and lived experience did not end when I physically left Malaysia. Just as the collapsing of these lived dichotomies was to be a constant and continuous challenge for me in the field, so too has it been a key duality I have had to confront *once back from the field* in the writing of the text, the representation of the experience, the telling of the story. For connected to this experience of disconnection was a question about knowledge and science as an armamentarium of practical applications and prescriptions, and the location of knowledge as a reflective mirror and an alternative reconstruction of reality. Indeed, as a politicized statement about the *lack of alterity* itself.

Central then to this struggle and this confrontation is a politics of representation that echoes the questions posed by the project itself. How should or can I represent the interconnectedness of these stories? What should I highlight? Which should I privilege? How to recreate in the telling of the story and the experience, the immediacy or situatedness of the ethnographic encounter? How to convey through concepts and words the physical and social moment that is now no more? And do so without diluting the complexity or richness of both field and experience? Can I do it and still claim an ethnographic or scientific authority that the story itself will likely declaim? On what other basis must I then place my ethnographic authority? My voice and my experience? And where is the Other in relation to this?

Through these questions, partially informed by my theoretical frame, I am compelled to return to the lens framing the project, and see in it a mirror reflecting my struggles back to me. Unmediated and unmitigated by the so-called data I had collected in the field, I had become part of my own subject and object of study in the most personal of ways. What had I seen? How had I seen? Who am I in relation to my objects of inquiry, and my subject matter? Indeed, what is it to think of Malaysia with both capital "M" and little "m" as an object *for* inquiry? How is this related or unrelated to the academic exercise of writing and knowledge construction?

A commentary on what it means to be a professional academic, the writing of the text became a struggle: against writing convention, against form, against structure—a veritable series of inherited and learned conventions ranging from the practical [how to write, what to write, where to write] to the epistemic [disciplinary boundaries, conceptual filters, beliefs about science, authority, contribution]. A process of ***unlearning*** had to be made, where notions about what is appropriate and necessary had to be deconstructed cognitively, textually, theoretically; and the theoretical armamentarium of critiques informing the current project, *appropriated* and *used* as a living experiment and reflection on lived experience.

THE PROJECT AND ITS QUESTIONS

Conceived in response to an amalgamation of factors that are theoretical, professional and personal, this project is an intersection of multiple possibilities and opportunities. Theoretically, it was formulated in response to certain questions raised by writings in anthropology, social theory and postcolonial studies highlighting the "situatedness" (Haraway, 1988) of modern disciplinary knowledge (e.g., development studies, science, anthropology, history), and their relationship to other fields of know how not traditionally considered by their respective disciplines. To writers raising this specter of a non-neutral and non-universal science, "knowledge" that we take for granted—for example, that development is an antidote to poverty and a barometer of progress; that history is a linear record of social or economic evolution; that science and the scientific method are universal ways of knowing—are first and foremost symptomatic and representative of a very particular form of knowing and living which is specific to a certain culture and time: Modernity (e.g., Chatterjee, 1993, 1987; Clifford, 1988; Escobar, 1994a, 1994b; Haraway, 1988; Harding, 1998; Prakash, 1990; Radhakrishnan, 1996, 1994; Rich, 1984; Said, 1989, 1979). These scholars pinpoint attention to the epistemic and cultural heritage of modern knowledge, the practices organizing the production of such knowledge, and the relationship between knowers (e.g., western self, researcher self), and known (e.g., non-western others, objects of inquiry) which underwrite the construction and representation of these knowledge.

In contrast to these views, the increasing prominence of, and interest in management and organizational research of an international and cross-cultural nature since the early 1990s also served as an entry-point for the project. During this time, calls for greater engagement by management scholars in international or cross-cultural research were fielded out across the academy, part and parcel of a generalized interest in globalization and its processes (e.g., Boyacigiller and Adler, 1991; Doktor, Tung and Von Glinow, 1991a). Triggered partially by the collapse of the Eastern European bloc and the economic success of newly industrializing nations in East Asia (e.g., Korea, Hong Kong, Singapore), this interest in "the global" often translated into an explicit concern with identifying or assessing how such emerging economies and nations may benefit from, or have benefited from, the introduction of free-market policies to organization and development.

Expressed more generally, scholars in the field were—and still are—concerned with learning how these economies would benefit from transfers of western management and organization knowledge. Most, albeit not all, of these scholars saw in the transitions and changes occurring in these economies, *a signifier for globalization itself.* To them, such transitions and changes were taken as evidence of the success, the efficacy and more importantly, the applicability and appropriateness of western forms of organization and administration to any and all contexts. In contrast to the growing recognition about the situatedness of modern know how that was emerging in the social sciences, an explicit and implicit "universal-applicability" was/is attached to the international and the cross-cultural in the management sciences. Hence, the aforementioned interest in, and presumption about, the benefits to be had from a systematic adoption of market-oriented policies by newly emerging or industrializing nations. It is in response to these disjunctures in theoretical and academic circles regarding what constitutes "the global" and "global knowledge" that this project is formulated around.

According to normative approaches in the field of international management, knowledge transfer as a process is composed primarily of the literal and linear transmission of specific [organizational, economic] procedures, skills, tools, practices and structures from the industrialized West to the Rest (e.g., Israel, 1987; Kiggundu, 1989). Within this context, research on transfers tended to focus on a variety of technically-oriented issues surrounding the mechanics and form of transfer: questions range from a concern over channels of transmission and communication to the scope and magnitude of know how targeted for transfer to interests in the directionality of transfers (e.g., Goshal and Bartlett, 1989, 1988; Gupta and Govindarajan, 1991; Osborn and Hagedoorn, 1997; Park and Ungson, 1997; Porter, 1986; Teece, 1976). Where culture is attended to, it is seen as either an aid or an obstacle to the transfer process (e.g., Gupta and Govindarajan, 1991; Osborn and Hagedoorn, 1997; Park and Ungson, 1997).

The approach I take in this project is one that sees this interest in transfers as *already situated in the culture of Modernity.* In this instance, it is precisely the modernist assumption regarding the universality of their knowledge that makes possible these researchers' unquestioning acceptance about the applicability of their know how, and the possibility of researching transfers as a singularly literal and linear activity of transmission. By contrast, my approach allows for exploring how the location of transfers as a modernist form of knowing and doing is informed by specific interconnections between culture, knowledge and practice. Attention is focused instead on how these connections are making possible the phenomenon of transfer as we know it, and what these connections imply about the construction of knowledge in the international and cross-cultural arena. Malaysia is, in this sense, a theoretical and empirical space for investigating the myriad of representational, social and methodological issues arising from this situatedness and these interconnections.

The interdisciplinary management scholar in me wanted to know: what can alternative approaches to knowing forwarded by scholars from elsewhere in the social sciences tell us about the processes of globalization? What can they tell us about the nature and forms of knowledge currently organizing transfers as a discourse, a practice and an activity? What are the everyday practices organizing it as a form of knowledge and practice? To this end, what if transfers occur in neither a cultural nor social vacuum? Nor are their prescriptions universal (i.e., culture-free). What if the assumed flow or movement of this knowledge from the West to the Rest is a non-linear one, filtered by processes of constitution and appropriation (e.g., Chatterjee, 1993, 1987; Ong, 1987)?

How would knowledge transferred and acquired under such circumstances look? What might be involved in the very process of transfer itself? Put another way, assuming the existence of an already cultured Other, what happens at such points where cultures "intersect"? And lest we forget, what of the Other? What does it mean to be a recipient of such know how? What does the acquisition and use of such transferred knowledge imply about changes to the local situation and the local self? Where and how is culture in all this?

Furthermore, if we are to take seriously the problematizations forwarded by social science scholars about knowledge and its conditions of production, what does this mean in terms of the research and representation process? What must the researcher be attentive to in such a context? Indeed, what would researching itself entail? What might now be part of the story that was not there before? How would or should this story be re-presented?

The researcher enters the field already defined by a relation (e.g., research, visitor, returnee) to his/her subject (e.g., development, management) and object of knowledge (e.g., Malaysia). Whether newcomer or returnee, what the researcher makes visible and documents is as much a commentary on the researcher and his/her discipline as it is the object of inquiry. That is, "where knowledge is coming from" and "whose knowledge it is" matters. By extension, this knowledge which is coming from somewhere by someone becomes an entity somewhat less "global" than it is otherwise assumed to be.

Given the above, the writer and ethnographer in me wanted to know: can I represent and offer an alternative way to understanding, seeing and presenting the objects of my inquiry which will allow for a richer and more diversely- or historically-sensitive way of understanding, seeing and presenting knowledge itself? As I note in passing at the beginning of this chapter, even within the otherwise modern spaces of foreign multinationals, a certain locality mediates understanding and experience of such spaces. How are we to understand these seemingly cultured mediations and negotiations? Are they to be dismissed as aberrations or designated as obstacles? Or can they be re-presented as an opportunity for us to understand how such cultured engagements may have something to say not only about who or how they are, but who and how we are? It is this *story behind the story* that is at the center of my project.

On a personal level, the decision to locate my study in Malaysia is very much connected to a promise I had made myself a long time ago—not exactly the one I made when I first left home, but connected to it nonetheless. Namely, that someday, I would return and "tell a story" about my relationship to this place I used to call home. But the project itself is not just a personal journey or an act of taking-stock; it is also political in intent. Returning with an awareness and a knowledge that the Malaysia of both memory and childhood extends far beyond the reaches of memory and childhood, I return to witness it not just as a home, but as a political and social site where history and the relation between a West and the Rest is lived and enacted on a daily basis.

The non-child in me who is revisiting home wanted to know: how is it connected to the West? In what ways are they connected? What is done to home in the interest of economic development and economic exigency? In the name of science and efficiency? And what can home tell us about this development and exigency, this science and efficiency? What, in short, is the history that is organizing this connection, this exigency and this knowing?

"WRITING-IN" SILENCE

Where is Malaysia in all this? As I was to realize time and again throughout the course of the project, the dominance of the aforementioned technocratic rationality, interest in functional and technical parameters of knowledge acquisition and use, focus on institution building and efficiency, productivity and performance—whether by dint of my own research focus or training, managers' identification of problems, or the country's preoccupation with progress—were constantly coloring what I, as a participant, observer and researcher was able to see, document, assess and experience. So much so that this *other* Malaysia—the one marked by communities adjusting to social and economic transformations, traffic jams, pink city buses bursting with passengers and absence of cows—was constantly in danger of being eclipsed by the immediacy of the research focus and national vision. This silencing of the social and personal by the official was, and remains, a commentary on the power of professional norms to define our sight; and an example of how apprehensions of reality are bounded by cultured conventions regarding desirability, process and science which extends *far beyond* the immediacy of the local situation.

When concern deals with and language is constituted by—objectifying and universalizing questions about economic production, administrative efficiency and so forth, the living presence of Malaysia *disappears*. In the process, what is silenced are the multiple layers of intersecting and dissecting factors denoting location and experiences such as race, gender, class, religion and history: signifiers of both country and choices made, whether personal, national or global.

The power of such linguistic forms and professional conventions to define the terrain and visibilities of the researcher as an "eye" of society and social experience came to my awareness time and again in a myriad of small and significant ways. About two-thirds of the way into my research, on the way home from one of my visits to the multinational in Penang, the taxi driver started a less-than-welcomed conversation. Feeling restless and distracted, I half-listened to his seemingly unending chatter about this and that occurring in Penang. Looking somewhat disinterestedly at the surroundings flashing by us—the massively impressive facades of the factories, the rumble of a dilapidated fume-spewing factory bus carting yet more operators to work, the high wire fence surrounding and protecting the boundaries of the Bayan Lepas Free Trade Zone—I wondered to myself: *is this it? Is this why I'm here? What I scrambled so long and so hard for?* Yet I would not have been able to articulate in any detail the reasons behind my distraction and my distance should anyone have asked; save to say that there was a disquieting sense of how familiar everything had come to be in just a few short months.

Forcibly turning my attention back to the taxi driver, I tried to catch what he was saying; a process of participation which eventually led to a back-and-forth discussion about changes taking place on the island, about certain dislocations of communities, construction accidents, landslides and the growing encroachment of resort-like structures—*proposed on Penang Hill no less!* [as he was quick to exclaim]—geared for consumption by tourists, both local and global [*what do they care? . . tsk tsk . . as long as they can make money it's okay, everything is money*]. I wondered for a split second if I was included in this "they" that he was talking about . . .

I realized in the interim that I had been so preoccupied with the exigencies of quality management, time optimizations, production runs and shortages of labor over the past few weeks—as I went about documenting the specific configuration of problems and practices confronting both companies and government alike—that I was in danger of forgetting that Malaysia as a community of people is *more* than the technocratic experts I had been interacting with: indeed, that these technocrats *themselves* are more than their technocratic expertise. It was a reminder that the professional languages used by experts are more than an economically-linked index of some neutral science and technology; that the official language and concerns I use in my conversation with these individuals and vice versa is only a partial—albeit an important one but still a partial—representation of the story on hand. The question *"where is Malaysia in all this?"* popped into my mind and I was confronted with the unpleasant task of having to recognize that I was in danger of being "gobbled-up" by the technocratic world I was traversing.

Falling slowly silent, I looked at the surroundings and images flashing by us as we continued to make our way home: the glimpse of a banana tree peeking through the upper boundaries of a walled-in garden; two schoolgirls chatting animatedly to each other, clad in the ubiquitous turquoise of their uniforms, bulging knapsacks hanging from their shoulders; the colonial facade of a house. And there, amidst the overhang from the gnarled old trees lining the boulevard leading me home—signature imagery of Penang itself—is an old man pedaling his *beca* [trishaw], clad in shorts and white t-shirt. I wondered where he had been and where he was heading to—a certain solitude marked his journey. What do these images and these individuals have to do with all the concern over calibrations, head count and training in this other world I had just left behind? How are they related? Linked?

And I realized anew that there *is little or no room* within this development and management language of economics and institution building for old men on *becas* and schoolgirls walking home, since neither are theoretical components nor research subjects of relevance within this language—at least not directly. *It is merely that simple.* And so what I attempt to do here is to "write-in" the *beca,* the knapsacks, the old man and the schoolgirls, so to speak: images of a Malaysia that can otherwise only be voiced in terms of economic parameters and organizational contingencies [or is it anomalies?] within the official languages of development and management.

More specifically, as I seek to tell in this story, understanding the forces propelling development in Malaysia is not just about referring to the particular specifications of its development policies. It is equally about understanding the confluence of historical, epistemic, social, economic and political relations compelling and underwriting these development policies. As it is about understanding the cultured and institutionalized lens by which this knowledge of development itself is gleaned and deployed globally, where histories and stories about the Third World are also the history and story of the West and vice versa.

In the context of Malaysia, as an example, this story is about understanding the importance of that "immense national sacrifice" known as the New Economic Policy (1970–1990). Important here not so much for its stated economic objectives and strategies, but for what it says about the scars in the nation's psyche and the subsequent actions taken to redress these scars; a consequence of how a nation that thought itself immune to communal strife learned otherwise.

In the context of this history, the story to be told is about understanding how it is that Economics—whether as national objective or as a concept central to modern notions of development[3]—is the spatial, epistemic, productive terrain and social, political, economic equalizer currently relied on by the state to construct a "common ground.'[4] Indeed, understanding how it is actively used as a national unification strategy that will cross boundaries and link all within its jurisdiction: a living embodiment of what it means "to nation" in the context of a history which remains rife with colonial and modern inequities. At heart, this work is about how nations like Malaysia—in negotiating for a voice in the global economic arena—must necessarily do so within the social, economic and political lacunae inherited as a result of its location as a former colony, and use a language and forms of practice not its own in order to be heard. Where the local arena, it would seem, comes already embedded within a global knowledge.

In short, this project looks at what it takes for so-called newly industrializing countries to be players in a field not of their own making; in a game entailing the active attendance and intervention by the state, and active participation by global capital into the pre-existing social, economic and political fabric of these societies—what Escobar (1994a: 213) had called the massive reconstitution to the "life," "labor" and "language" of Third World societies by western technologies, science and development.

And so I am in Malaysia once more to tell a story and witness [or is it to participate in?] an history in the making: a sometimes circuitous trek in time, and an all-time account of the complicity between power and knowledge—at times visible, at times not. It is a story and a history which all of us as constructors of knowledge about Others, whether anthropologist, historian, development expert, management scholar or manager, have a role to play in its production and constitution. In short, it is about the continuing saga of worlds in collision, cultures in commotion, and identities in transition: where *right is the power of voice, and power the right to voice.*

Thus, like the reality it attempts to reflect and represent, this text is a battleground of multiple words and multiple worlds. It is a war of rights and representation, of power and voice, where lines, drawn to demarcate boundaries, are never clear-cut, but blurry and constantly cris-crossing and undercutting each other; where knowledge about Others become instead, an unending and impossible journey of "self-discovery";[5] and globalism a state in permanent co-existence with the construction of boundaries and nationhood. Malaysia is—metaphorically speaking—a world at war: the researcher [or manager, government official, and development expert for that matter] simultaneously a battleground, a lens and a weapon by and through which this war is fought.

SUMMARY AND OVERVIEW OF THE BOOK

This book is an attempt to confront a multiplicity of issues and practices—theoretical, methodological, professional, procedural—governing the constitution and construction of management knowledge in the international arena. An active meditation on questions about representation and voice, the phenomenon of knowledge transfers in Malaysia serves as an entry-point for investigating, locating and reflecting on these issues. The central challenge of the project lies, therefore, not only in documenting or understanding the current strategies, activities and so forth informing Malaysia's development initiatives, and the role that international management knowledge plays in this effort. It lies equally in the active *creation of a space*—theoretical, methodological, textual—by which silences and gaps produced by modern discourses and practices could be rendered visible and understandable as that very "emptiness" which is *actually* constituting relations between the official and unofficial story. The intent and challenge is to render "palpable" the presence of this silence so that it too becomes part of the text and part of the story.

At the center of this possibility are a conceptual frame and a methodology that call our attention to the cultured practices and relations of knowledge and power making up and making possible our ability to know others. It is an understanding and a possibility that open up for consideration a myriad of visibilities, assumptions, procedures and inscriptions attending the very *process* of cross-cultural knowledge construction. As I was to realize soon after my arrival in the field, locating something called "Malaysia[n]" is first of all, well-nigh impossible. Which Malaysia am I talking about—the official or the unofficial one? Whose experience can I talk about, whose should I privilege? For that matter, does it and will it matter who I talk to or listen to in trying to tell this story about the transition of a nation and the challenges and transformations ensued by it? What about culture? Who has it? How to identify it? Is it even clearly identifiable? Or is it, as some post-colonial writers contend, constantly in the process of being articulated? Where is the researcher in this process of identification and articulation? What role does he or she play in their constitution? Does it and will it matter where I come from and how I see? Moreover, if culture as a sense-making filter is necessary for someone to identify [another] culture, what influence do such lenses have on who or what and how we identify? What about the filter of professional training? These questions that I ask constitute the [representational] terrain making up my project and the story as I seek to tell here, and in chapters to come.

In chapter two, I introduce the methodologies of institutional ethnography and multivocality. I focus on the central premises of institutional ethnography, and pinpoint its significance to the project and its aims. More an approach to phenomenon than a method per se, institutional ethnography's attention to those procedures and relations (e.g., discursive, professional) that are constantly framing or organizing how and what we know provides the necessary entry-point for investigating social phenomenon which holds as central the various interconnections between knowledge and power noted previously. It is a centrality that makes possible my/our ability to see all participants to a phenomenon in terms of their respective locations as knowers and known, important here given the focus on representation and voice, knowledge and power.

Concomitantly, multivocality is a way of responding to the problematizations about knowledge, knowing and practices raised by institutional ethnography. In line with the premises of the latter, multivocality does not take for granted the objects and subjects of inquiry (e.g., Malaysia, conveyers of knowledge, recipient targets). Instead, it sees in these objects and subjects, a multiplicity of *relations*—historical and current, epistemic and institutional, economic and location—that characterize their connections as knowers and known, and seeks to reconstitute how phenomenon can be re-presented and made understandable from this standpoint of relations and location. Through such confrontations with how knowing/knowledge is organized and rendered knowable, both methodologies call into question the impossibility of having *or producing* knowledge which is *not* located or situated in one way or another (i.e., culture-free, universal).

Chapter three looks at the official "gaze" of the international and cross-cultural management literature. Part of the institutional ethnography making-up this project, it draws on insights offered by recent deconstructions of the development sciences and postcolonial historiography to anthropologize this literature. Situated within the context of Modernity, specific attention is focused on the concepts, assumptions, practices and constructions of subjects forming and informing this literature as a field of knowledge. In doing so, it pinpoints attention to the multiple factors and relations (e.g., cultural, discursive, professional, organizational) framing the visibilities and actions of scholars writing on the international and cross-cultural arena. This review and analysis provide the epistemic and theoretical framing needed to situate and locate the actions of managers and officials in chapters to come.

Chapter four focuses attention on Malaysia: an account of the multiple intersections between the local and global which had gone into the constitution of the nation and its development policies, the chapter looks specifically at how neoliberal discourses of development are currently changing the state's constitution of its development problems and solutions. More specifically, by attending to the engagements between global discourses of development and local concerns with development, the chapter provides an entry-point for situating the nation's current interest in transfers of management knowledge.

Picking up where the previous chapter ends, chapter five considers in some detail the kinds of knowledge and practices that are being put into place by multinationals and state planners alike in the name of development and transfers. Focusing specifically on the concept and practice of Total Quality Management (TQM), the material presented here comes from the multiple institutions I examined in Malaysia. By assessing the constructions of subjectivities forwarded by managers and officials in their bid to "transfer," attention is directed to the concepts, aims and relations making possible this interest in all things quality, and the subsequent changes to "life, labor and language" that are set into play as a result of this focus on quality as a barometer of progress, and antidote for underdevelopment.

Chapter six re-tells this process of constitution and deployment from the standpoint of culture and its mediatory effects on both the local and global. On the one hand, it highlights how the transfer process is informed by a need on the part of local managers and officials to actively mediate or translate "culture" in their location as bridges between the local and global. On the other hand, it shows how this process of translation is always already bounded by a local—in itself, a process of mediation and appropriation that changes in both meaning and form the very knowledge and practices so deployed and/or adopted.

Chapter seven focuses on the disjunctures, counter-representations and negotiations set into play as a result of changes [procedural, technical, institutional, cultural] being affected in the nation. Attention to the nature and form of these contestations and counter-representations highlight the mediatory power of the local, while also directing our attention to the sites and affects of power attached to knowledge and practices being put into place. By constantly referring these negotiations and contestations back in terms of the local and global particularities that were/are organizing locals' speech and actions, the chapter highlights the often limiting and paradoxical nature of modern management and development knowledge that are at the center of these contestations.

Through their attendance to the local and counter-constitutive, chapters six and seven are attempts on my part to remain accountable to the agency of subjects who are simultaneously owners of cultures and histories all their own, and knowing agents of both their lives and experiences. Constant attention to the presence and location of the researcher "I/eye" in relation to my subjects throughout this process of being attentive to the local focus attention to the often dialogical and relational ways that knowledge itself is negotiated into being. In effect, the multivocality indexed by these different ways of *acknowledging* and *writing-in* the local, the disjunctive and the relational are directly connected to this project's aim of looking at *how else* "the global" and globalization may be understood.

In chapters four, five and seven, an unofficial story is interwoven and inter-spliced into the official text. Much like what I did at the beginning and middle of this chapter, these unofficial stories provide a sometimes complementary and sometimes paradoxical adjunct or juxtaposition to the official ones embodied by the main texts. They were selected and written-in to highlight specific features, aspects and/or components attached to changes currently occurring or that had occurred in the nation that are *otherwise silent* in official accounts of the nation in transition. Intended as a series of [personalized] commentaries on the official story, they focus our attention to the multiple possibilities, meanings, effects and/or alterities making up the social terrain of the nation and the polity as it is lived on an everyday basis.

Finally, chapter eight situates the official and unofficial stories told previously in terms of an alternative possibility for understanding knowledge transfers. The chapter calls attention to the multiplicities and shifts in meanings that attend the transfer process, and shows how the story of transfers as told in preceding chapters, contravenes and challenges the presumption that managing or researching in the international or cross-cultural arena is reducible to a question about the knowability and predictability of Others. In doing so, this "re-telling" of the story highlights issues governing research process and knowledge, representation and voice connected to mainstream understandings about knowledge transfers. How, in effect, does one research phenomena which are constantly evolving in form and feature, composed simultaneously of processes of constitution and contestation, involving people and institutions who are neither passive nor empty of meanings? I discuss these issues and challenges in terms of how the approach and methodologies I forwarded and used have sought to address and redress them.

Chapter Two

Institutional Ethnography and Multivocality

This chapter introduces and reviews the specific features of institutional ethnography and links it to the book's interest in issues of representation and voice, as well as silence and *re*-presentation. Its link to the latter is closely connected to the reconstructive aims of the project, articulated here via the focus on multivocality. I start by exploring the central claims of institutional ethnography before proceeding to consider its significance. Specific references to the research process in Malaysia and the strategic or methodological reasons grounding the formulation of its chapters are used to locate and situate the project in terms of the requirements of institutional ethnography itself. In the final segment, I consider the requirements for a more locally and historically sensitive approach to knowledge, and its significance in terms of the knowledge construction process.

INSTITUTIONAL ETHNOGRAPHY

A Way to Problematize and Situate Knowledge/Knowing

Institutional ethnography (Smith, 1990, 1987) is not a "method" (e.g., observations, surveys) in the conventional sense. Rather, it is a methodology and *an approach* to studying phenomena. As a methodology, institutional ethnography does not take for granted the phenomenon under inquiry, looking instead at the phenomenon *as* an object and subject *for* inquiry. Neither does it accept as truth existing knowledge or understandings about the phenomenon. This knowledge becomes instead, an object for investigation. Namely, it seeks to understand phenomena by first asking *what is making it possible* for us [scholars, researchers, managers] to understand phenomena in the way that we do.

What are the concepts informing our sight? What are the representational practices and assumptions that make this possible? What in turn, does this knowledge prescribe? Who are the objects of its knowledge? Indeed, what are the social, political, economic and academic or organizational factors organizing this interest in the phenomenon? In effect, the methodology asks the researcher to investigate and situate the social particularities circulating around and informing the phenomenon as an *entry-point* to understanding the phenomenon itself.

To institutional ethnography, this entry-point of focusing on existing knowledge is important because knowledge is always and *already* part and parcel of how and why we understand a phenomenon in the way we do. To the extent that this is so, further attempts to investigate its features should be actively attentive to this knowledge and the concepts, assumptions and practices that informed the production of this knowledge in order to "make visible" the conditions and limits that are attached to it. From the standpoint of the methodology, to do otherwise is to risk perpetuating possible limitations attached to what is already known, or to remain unaccountable for how this knowing was configured into being. In the context of this project's interest, an institutional ethnography of knowledge transfers means starting out by *not* taking for granted what is often accepted or known about transfers: for example, the belief that our knowledge is universal and transfers a literal or linear process of transmission.

Smith (1987) frames this recognition and necessity in terms of "the problematic of the everyday," and the "conceptual practices of power" (Smith, 1990). According to her, the everyday world and our knowledge of this world are *already* problematic because what we see and assume to be True (e.g., transfers as literal transmission from self to other) is already the product or result of certain histories, beliefs about knowledge and/or practices which are "*not* fully apparent . . . or contained in" this everyday world (Smith, 1987: 92, emphasis mine). Its problematic nature arises from the *invisible* and *silent* way these beliefs and so on are constantly framing and organizing our thoughts and actions in our encounters with events, with people, with knowledge and so on.

As an example, the systematic construction and reliance on concepts, categories and standards that scholars and researchers create and depend on in order to know (e.g., individualism, collectivism) is a central procedure by which the everyday is socially organized. This is so because these concepts and categories are already the result of certain "ordering procedures" (e.g., objectifications of reality, assumptions about science) "established in the theoretical armamentarium of the discourse(s)" they constitute and/or are constituted by (Smith, 1987: 122, emphasis mine). Accordingly, when we know or see through these concepts, the "facticity" of the reality they present to us is one already mediated by this plethora of ordering procedures *which are now invisible* to both the act of knowing or seeing, *and* the knower or observer. Thus, what the scholar or manager takes to be normal everyday activities (e.g., researching, planning, managing) are activities already extra-locally framed by these practices and knowledge. To Smith, it is this silence, and this disconnection between thought and action, profession and activity, or knowledge and practice, which make the everyday problematic.

Institutional ethnography is, in this sense, somewhat akin to ethnomethodology (e.g., Garfinkel, 1967) in calling the ordinary reality of institutions into question, and symbolic interactionism (e.g., Goffman, 1959) in its focus on the social construction of reality. Like ethnomethodology, institutional ethnography focuses on everyday practices, the production of the social, and the creation of order, albeit via a focus on concepts and categories in "accomplishing the orderliness and sense of local processes" (Smith, 1987: 161). Unlike symbolic interactionism however, institutional ethnography holds as central the focus on a problematic everyday as an *entry-point* for evaluating the social construction of reality. Thus, while definitions of situations and the rules sustaining them are important, knowledge of how these rules are inscribed into activities and come already socially organized [i.e., framed extra-locally] are equally important. As Smith notes (1987: 97), by considering the everyday as a self-contained and comprehensive phenomenon, symbolic interactionists "seal off" the everyday from its "sociological universe."

Accordingly, the aim and method of institutional ethnography is to constantly refer our attention *back* to what might be bounding or shaping our constitutions of reality and phenomenon, including as an example, the previously noted ordering procedures making possible the existence of concepts, our reliance on concepts and so forth. By doing so, the methodology provides a way to simultaneously problematize and re-situate the phenomenon under inquiry, our knowledge of the phenomenon, and those social particularities that are organizing what we take for granted about it: what it is, how it is, why it is.

In the case of transfers in Malaysia, of particular significance for knowing what to attend to in such an institutional ethnography are insights from recent deconstructions of the development sciences and postcolonial historiography (e.g., Chatterjee, 1987; Crush, 1995; Escobar, 1994a, 1994b, 1988; Manzo, 1991; Mueller, 1987a; Prakash, 1990; Radhakrishnan, 1996, 1994; Sachs, 1992; Shiva, 1991, 1989). These writings are significant since these scholars already question in their writings the nature, process and feature of how we know and what we do in the name of development in particular, and knowledge construction in general. As noted by Escobar (1994b) for example, the modernist nature of development discourses, and the connections between knowledge and practice underwriting them are particularly important social forces explaining the constitution [i.e., representing], as opposed to merely reflecting or documenting [i.e., presenting], the underdevelopment of the Third World to us.

This focus on interconnections and relations between knowledge and power are central to Smith's (1987) very constitution of institutional ethnography itself. The notion of institutional as used by her (Smith, 1987: 160) refers *not* to some "determinate form of social organization," but to a "complex of relations"—between researcher and subject, episteme and knowledge, knowledge and practice—which "intersect and coordinate" each other (e.g., concepts and sight, culture and understanding), and is "interpenetrated" by discourses of one kind or another to form the "ruling apparatus." Concurrently, ethnography is *less* a concern over methods than it is "an exploration, description and analysis of this complex of relations" (: 160). Put another way, looking at and exploring the multiple ways—discursive, scientific, professional, cultural—that relations are organized and intersect is, according to Smith, what makes a study ethnographic in nature, not the kinds of data collection methods used by a researcher.

Thus, an institutional ethnography of recent knowledge transfers might begin by directing our attention to its modernist heritage and to those interconnections making possible its constitution as a form of knowledge and practice. This I do in the next chapter in my analysis of the international and cross-cultural management literature. In undertaking such a historicized and situated review, the chapter highlights the already problematic nature of our knowledge on and about others; namely, cultured, organization-centered, reductive and economized. Equally, a benefit accruing from such an analysis is the possibility of recognizing the problem of undertaking such a review *in the absence of* any reference to those very forms of knowledge and power framing and organizing the construction and deployment of this [or any] knowledge to begin with. In effect, what this institutional ethnography of the literature or field of knowledge accomplishes is to highlight the *located* nature and feature of what is otherwise assumed to be natural, universal and/or culture-free.

This strategic intent to problematize and situate also applies to the organization of the other chapters in the text. In the instance of Malaysia, what is equally significant for understanding the interest in knowledge transfers are particular relations between the local and global, whether economic, historical and so on. Accordingly, an institutional ethnography of knowledge transfers in the nation would require attentiveness to these forms of engagement. As the historicized reconstruction of Malaysia presented in chapter four shows, there are a myriad of social, economic and discursive intersections between the local and global that are organizing action and intention in the nation. Indeed, these intersections are key for locating the nation's interests in transfers, and what is put into place in the name of this interest. As chapter five showcases, interests in TQM [as concept and practice] have to be referenced and situated within the conceptual, representational and historicized frame that such intersections effectively are.

Meanwhile, the numerous cultural, dialogical, institutional and procedural engagements and mediations surrounding and organized by this deployment of quality and other related forms of know how presented in chapters six and seven focus attention to yet another series of interconnected and simultaneously occurring processes that attend the very activity of transfer itself: for example, engagements between the researcher and subject, subjects and their objects of knowledge and so on. Together, these chapters tell a story about the multiple relations, forms and connections that are all *intersecting* to make possible our focus on Malaysia as a site and object of knowledge transfers [see Figure 2.1].

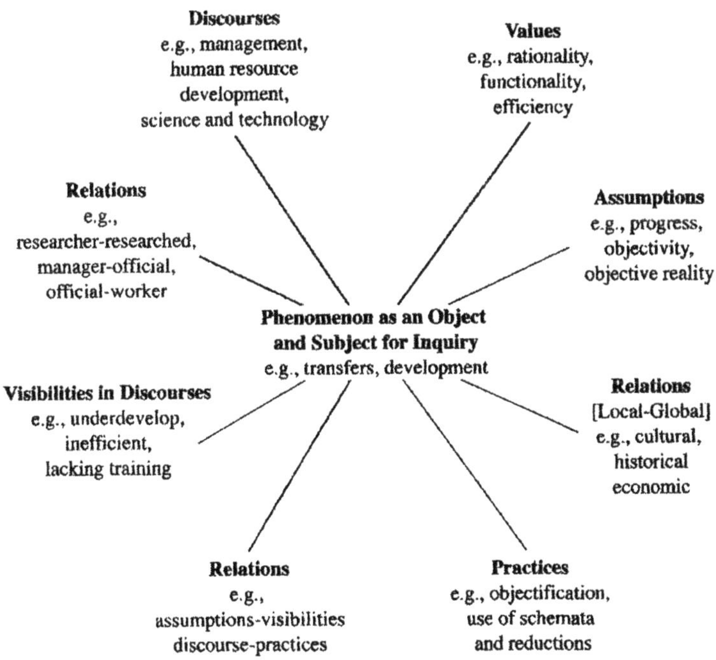

Figure 2.1 The Methodology of Institutional Ethnography

Thus, the task of the researcher is to identify, configure and re-configure what and how these central organizers and conditions of possibility—discursive, conceptual, procedural, professional, relational—are making possible the existence and activity that is the phenomenon of interest. While each chapter and aspect of the phenomenon identified earlier may or can be looked at separately, their very connectivity or intersection—made possible by the existence of a few basic organizing assumptions, concepts and practices, and underwritten by very specific relationships between participants in the phenomenon (e.g., manager and official, worker and manager)—is what institutional ethnography seeks to help the researcher to "un-piece" and "re-piece."

Significance of Institutional Ethnography

These simultaneous processes of questioning and re-situating are two key reasons why institutional ethnography is significant. From the standpoint of representation and voice, this questioning and re-situating are ways of compelling the researcher to question what he/she takes for granted, whether in reference to knowledge, its claims to truth, practices and objects or subjects of inquiry. Additionally, they are ways of being attentive to, and taking accountability over [indeed, of *remaining* accountable to] the social conditions giving rise to, and produced by knowledge (Radhakrishnan, 1994).

More specifically, given the central role that knowledge plays in configuring everyday speech and action, institutional ethnography actively incorporates accountability as part of its methodology by compelling researchers to focus on their effective locations as agents of knowledge. Notably, as constructors of knowledge, the lens and practices that researchers use—or bring with them into the field—are crucial in shaping and affecting the terrain and forms of visibility informing action and sight. To the extent that this is so, issues pertaining to representation and voice take center stage.

This focus on knowledge and power does not, however, apply only to researchers and knowledge constructors. It is equally the case for all participants to a phenomenon, be they managers, government officials, development experts, recipients of transfers, and so on. To institutional ethnography, the issue is not so much the participants per se, whether researcher or subject. Rather, it is the knowledge—and the various entry-points or relations *to* knowledge—which is socially organizing all these participants in their engagements with self and other that is crucial. Accordingly, investigating the various entry-points and locations—discursive, cultural, historical, professional or epistemic—that people "come from" is what is of central interest to the methodology; as are ways that these entry-points are mediating relationships between participants (e.g., recipients and disseminators of knowledge), as well as between participants and phenomenon (e.g., recipients' understanding of knowledge transfer).

This re-constitution and re-location of participants as active owners, subjects and/or agents of knowledge is what institutional ethnography aims to highlight with its focus on the problematic of knowledge and the reminder that what is important is not the phenomenon, but what is organizing the phenomenon and the relations between participants in this phenomenon. In the process, what it creates is a conceptual and methodological space by which to hold knowledge constructors accountable to their knowledge, while giving subjects their rightful space as owners and agents of their own voice.

These features of institutional ethnography are particularly significant in the context of studying the Other because it is a way of knowing which leaves the objects and subjects of inquiry "alone." As noted by Smith (1987), the object of the methodology is to leave subjects alone rather than reconstituting them into [passive] *objects* of the researcher's attention *and/or* knowledge. To quote her, " institutional ethnography as a way of investigating the problematic of the everyday world *does not involve* substituting the analysis, the perspectives and views of subjects, for the investigation by the sociologist [ditto management scholars and so on]" (Smith, 1987: 160–161, emphasis mine). This constitution of the relationship between the researcher to his/her subject of knowledge is a condition made possible precisely because of the methodology's interest in what is organizing and shaping the actions and speech of these subjects and objects of inquiry (e.g., beliefs about transfers, assumptions about progress) rather than the subjects or objects per se (e.g., transfers, Malaysia, Malaysians).

It is a way of knowing that comes with the presupposition of *not knowing—possibly even never knowing*—however contradictory or paradoxical this may seem. In the specific instance of neoliberal development and knowledge transfers, categories of analysis such as "market friend[liness]," "high shares of international trade," "strong competition" and "heavy investment in people" used to locate and explain the economic success of nations like Malaysia (e.g., Edwards, 1995; World Bank, 1993a, 1993b) are seen *less* as a commentary on Malaysia or Malaysians than they are signifiers for what we consider to be important. In a like manner, schemata such as those developed by Geert Hofstede (e.g., 1987, 1985), and other descriptors or categories like motivation, work values and commitment used so often by cross-cultural management researchers (e.g., Adler and Boyacigiller, 1991; Hofstede, 1987; Kedia and Bhagat, 1988; Ronen and Shenkar, 1985) to identify, categorize and label Others are seen less as knowing than they are systematic ways of organizing Others. Specifically, for organizing them *into* the "categories of knowing" which are singularly the analyst's or researcher's, not the locals." So that what is known is less "the Other" than it is a researcher or knowledge constructor Self who is engaged in the process or act of knowing.

This is particularly important in the case of this book, given the interest in knowledge transfers as an entry-point for locating and assessing issues of representation and voice. According to development and Third World critics, what we take to be the Underdevelopment of Others has more to say about what and how we know, the histories organizing relations between us and them, and our own interests in the "development" of these societies (e.g., Crush, 1995; Escobar, 1994b; Esteva, 1987; Mueller, 1987a; Parajuli, 1991; Rahnema, 1988a, 1988b; Sachs, 1992; Sheth, 1987; Shiva, 1991, 1989). To these scholars, development as we know it now has led to large-scale and debilitating consequences for most Third World nations. From this perspective of discursive and historical accountability, institutional ethnography is a way to *avoid* un-reflexively using as our starting points of analysis, those very knowledge and claims used so often as entry-points for organizing others and their histories.

The implicit approach being configured and articulated here is akin in certain ways to an almost ground-up-like approach. The object here is to let subjects tell their own stories. Rather than assuming *prima facie* the knowability of subjects, the researcher attends to how subjects' [re]-presentations of self, of events, of knowledge and so on, are articulating themselves *within* those discourses, institutional locations and relations—between researcher and subject, manager and worker, history and location—that may be or are organizing their speech and actions. In this way, the methodology does not assume *a priori* that others are [essentially] knowable, and confers instead a certain agency to them whether as subjects or agents of knowledge.

Accordingly, what gets called into question is the very assumption of knowability informing most traditional ethnographic and cross-cultural management endeavors and the representations these studies would present to us. Rather, what are seen are certain procedures and relations organizing the activity and practice of knowledge construction itself. Or, as alternatively forwarded by Silverman (1993), the "situated nature of knowledge generation" means not only that "no 'essential' reality exists" (: 201), what can only ever be known is the "practical activity of participants [researchers, subjects] in establishing a phenomenon-in-context" (: 203).

Thus, in contrast to the ethnographic tradition of interpretive anthropology (e.g., Geertz, 1983, 1973) for example, the intent of the methodology is *not* to gain "access to the conceptual world in which our subjects live" (Geertz, 1983: 24). Implicit within approaches such as this is the assumption of an underlying order of experience and meaning which can be systematically uncovered and categorized; an Other who is authentic and can be known or represented thus. Or as alternatively observed by Silverman (1993: 24) in his critique of most ethnographic studies in traditional anthropology, these studies rely on "the possibility of isolating and knowing subjects' interpretation of social reality," "ignoring" the social practices organizing situations and knowledge "in favor of perceptions."

To Smith (1987), this assumption of knowability and *privileging* of the researcher-ethnographer's perceptions [or reliance on experiential authority (Clifford, 1986)] are part and parcel of the ordering procedures and relations that produce—and hide—how and what we know. As I seek to show in my analysis of the cross-cultural management literature, the writings and representations of these scholars are filtered throughout by concepts and values such as motivation, efficiency and individualism specific to their field which make possible their knowing (e.g., Adler and Jelinek, 1986; Hofstede, 1987, 1985; Kedia and Bhagat, 1988; Negandhi, 1985; Ronen and Shenkar, 1985; Tayeb, 1987).

From the standpoint of researching the international and/or cross-cultural, this mediation of thought and action by such objectified and conceptual abstractions of a field are particularly problematic because they make possible researchers' ability to *apply* the general to the local such that the local is *removed* from its historical, cultural and social particularities (Smith, 1987). Seen from this purview, constitutions like the earlier description of the East Asian Miracle is less a commentary on the nations studied than it is a commentary on the World Bank "talking about" these nations, via the conceptual and documentary mediation of categories like "international trade" and "competition." To Smith (1987), such an extra-local framing of reality is nothing less than an "institutional production of reality" (Escobar, 1994a: 158) whose effect is "the privileging of certain subjectivities over others."

It is precisely with the intent of avoiding such a possibility that institutional ethnography calls our attention to the problematic of the everyday, and focuses this same attention to the framing of this everyday. Thus, not only is a conscious self-reflexivity "written-into" the research process, this concurrent awareness [pre-, in and après fieldwork] about existing conditions and limits is *precisely* what creates the conceptual space needed to remain accountable and open to alternative possibilities that may exist in how we can know, or what there is to know.

Thus, rather than assuming knowability, institutional ethnography is interested in *how* Others are rendered *knowable* and *known*. In this way, the methodology accounts for the procedures and claims of its knowledge while holding the researcher and other like knowers at the center of the plethora of relations and practices that are actively involved in organizing social reality and knowledge. This focus on the researcher and his/her practices recalls the writings of social anthropologists calling for a more self-aware and accountable epistemology (e.g., Clifford, 1988; Clifford and Marcus, 1986). As an example, Clifford (1988) has paid extensive attention to the rhetorical strategies and forms of institutional and methodological conventions used by professional ethnographers to legitimize their authority. By situating and writing-in the presence of the ethnographer and the ethnographic act, critics like Clifford not only bring our attention to what makes possible the ethnographic activity, they also bring into focus the presence of the ethnographer as an active agent and mediator of knowledge *and text*.

In effect, this attentiveness to the presence of the researcher and the forms or practices making possible knowing and knowledge is institutional ethnography's way of writing-in the possibility of recognizing that what and how we know [whether researcher or subject] is always *coming from* somewhere, and *speaking to* something. Concurrently, by leaving the objects and subjects of inquiry alone, the methodology remains attentive to the possibility of being open to the multiple stories that may or can be told by those we study. In this sense, institutional ethnography facilitates the project's aim of generating a more multi-focal and located approach to knowing and representing Others.

Finally, by calling our attention to the ways that knowledge and practices are constantly intersecting and bisecting each other, institutional ethnography provides a way of getting-around and mediating *between* the various dichotomies anchoring modernist constitutions of knowledge and reality. Included here are dualities like micro-macro, individual-institutional, past-present, local-global and knowers-known. As I seek to show in my analysis of Malaysia's interest in knowledge transfers, understanding the reasons for this interest entails a reference to certain interconnections and relations between the nation's history and present, between global discursive practices and local concerns, between the interests of global capital and official concerns over poverty which are *all coming together* and intersecting to render possible a certain constitution of problems, assessments and/or responses leading to and making understandable the nation's focus on transfers.

In this sense, the focus on intersections is a method of analysis which sees, as an example, that history is not something which is past and gone, but is an ever-present and lived experience, organized and mediated by memory, by location, by knowledge, by what others are doing, by what is presently going on and so forth. By bringing our attention to these intersecting and relational possibilities, Smith presents and provides to us, a way of capturing and representing the complex *simultaneity* and situatedness of our everyday worlds and phenomena.

RESEARCHING "MALAYSIA" AND RESEARCHING IN MALAYSIA

The terrain of Malaysia was one already steeped in a global language at the point of my arrival in the nation. The interest everywhere, or so it seemed initially, was centered on the exigencies connected to development, quality management, efficiency, productivity and the like. There was not much by way of an authentic field or native other, so to speak. The challenge for me involved trying to understand how the ostensibly global was *in* the local, and how or where "the local" was in the midst of all this. It was not until later, as I reacquainted myself with the nation, the concerns informing its policies, and the multiplicity of its local interests that my understanding of its decisions and choices, its nature of global embeddedness and so on was reconfigured.

Methodologically, I interviewed a myriad of individuals, including managers, officials, trainers and engineers; and observed a variety of events or situations such as meetings, presentations and interactions making-up the nexus of transfers I was interested in. As I did so, a number of questions aimed at locating the problematic nature of the everyday, and the intersections between culture, knowledge and practice which are/may be acting on the local were constantly kept at the back of my mind in order to ground my observation and analysis. When a manager or development official speaks, where and what is this speaking traceable to? Who is this person speaking as? What is organizing his/her speech and actions? More specifically, what are the assumptions and constructions embedding such speech or action? What are the practices and forms—conceptual, textual, documentary—constituting and mediating this speech and action? What kind of a language is the knowledge constituting and mediating these conversations and behavior? How and what is it connected to, both in and beyond the immediacy of the site and event? Where is it being used? When and in connection to what? Indeed, who speaks and who or what is the object of this speech and action?

Questions such as these serve as cues, and ways of attending to the multiplicity of interconnections organizing thought, action and intention at the local level. Written into the manner these questions are asked—informed by both frame and methodology—is a concerted attempt to situate the everyday into its larger social, material and historical underpinnings. In this way, the specific historical relations organizing the project (e.g., modernity, development, knowledge), and the nation [Malaysia as object or site of transfers] was implicitly and explicitly being accounted-for and written-in.

Fieldwork in Malaysia, undertaken throughout 1995, was located in the states of Penang and Selangor. A combination of observations, interviews and secondary sources from institutions[1] I visited such as policy objectives, annual reports, administrative guidelines, evaluation forms and internal bulletins' comprised my forms of data collection [see Appendix A for a list of these institutions and data collection particulars]. In total, fifteen institutions were tracked through for the study. Research sites included two multinationals in the electronics industry, and various government, statutory and academic institutions. The multinationals were used as starting points for my fieldwork because they are important spaces of Modernity (e.g., Escobar, 1994a, 1994b; Ong, 1987), and have traditionally been—and still are—understood to be key sites of technology and knowledge transfers (e.g., Blake and Walters, 1992; Gupta and Govindarajan, 1991). Finally, they are significant arenas of activity representing Malaysia's export industrialization strategy (e.g., Ali, 1992; Ismail, 1995; Ong, 1987).

Along with fieldnotes from interviews and observations, a series of reflective notes were taken to record the myriad of issues, concerns, tensions, feelings and so on locating my experience in the field. Half the time, these reflections were not directly linked to the sites I was observing, but to a more generalized experience and observation surrounding the focus of my fieldwork itself. Covering a broad range of sometimes loosely-connected issues and concerns including the everyday experience of living in Kuala Lumpur, researching in Malaysia, conversations with interviewees, friends, family, and yes, even harried and chatty cab drivers, as well as analysis of newspaper reports, these reflections served as an invaluable locator and guide to what living in Malaysia was like during this time and a reminder about the inter-connectedness of what I was experiencing as an individual and what I was documenting as a researcher.

My first concern as I set out on my fieldwork was to identify the major strands of management philosophies and techniques currently in use within the multinationals. This would provide information on the official focus of transfers and development activities, as well as an entry-point for tracing and assessing the intersections locating transfers as a phenomenon, including those epistemic, textual and administrative forms and practices constituting the local terrain. From the onset of fieldwork, the focus on quality, productivity and efficiency—mediated often through official policies and practices on Total Quality Management (TQM)—was to be a defining feature and focus of the institutions, activities, events and speech I was observing and documenting.

Observations and knowledge about the multiple elements making up this focus on quality and productivity cued my attention as I went along. Accordingly, much of my time at the multinationals was spent documenting and taking note of the particular constitutions of this knowledge, the documentary and mediatory forms it takes, when and how these were used, its connection to official goals, the agents or practices involved in its deployment and so on. Towards the end of fieldwork, a "profile" making up this focus on quality management was configured which included information on (1) the activities such as training events, (2) the administrative or documentary practices like planning, tracking or reliance on categories of evaluation, (3) the relations between institution to institution, institution and knowledge, knowledge and management, (4) the forms of knowledge, whether managerial, economic and/or technical, and (5) the conceptual or cultural filters such as efficiency, productivity and so on locating it as a discourse and an object of transfer.

Often, information garnered as I negotiated my way through these institutions provided further cues on relations—institutional, economic, epistemic—connecting this focus on quality within the companies *to* [*or from*] institutions *elsewhere*. For example, a television special on multinational-government joint ventures on training in the electronics industry was provided to me by one of the TQM staffers at Multitech. Typically, such identifications would come from the managers and officials during my interviews or informal chats with them. As another example, connections between the multinationals and certain statutory agencies like the National Productivity Center (NPC) was brought to my attention by Multitech's TQM training and development manager in her explanation about the company's award in quality excellence. Subsequent meetings with personnel from the NPC in turn, led to information about the linkages between the multinationals, the NPC and the Standards and Industrial Research Institute of Malaysia (SIRIM), along with information on parallel institutions that exist in and for the public sector. More often than not, suggestions as to where I should go and who I should see would be freely and generously given, including names and titles, each individual acting as a reference for the next to be approached.

What is interesting from the standpoint of institutional ethnography is the naturalness with which such information was conveyed. There was a certain casual-ness in how managers and officials would direct my attention to institutions and relations; an almost assumed obviousness attached to their delivery of this information. As in the case of the aforementioned manager at Multitech, her intent was not so much to convey information on the NPC as a connection [however this may be understood]; rather, information about this connection between the multinational and the NPC came *in the act* of her describing the company's "journey" from non-TQM organization to winner of the Prime Minister's Quality Award: connections, it would seem, were not only evident, they were natural. Reflecting back on this, the question arises as to what this may or could mean since connections so conveyed are also a commentary about how terrains [whether MNC or NPC] are socially organized by the intersections or relations constituting them, whether institutional, administrative or conceptual. Is such naturalness an index of the problematic nature of the everyday?

Throughout this time, I vacillated between an Alice-in-Wonderland-like curiosity to impressed admiration and absorption, to an almost claustrophobic reaction to all things "quality." The activities, concepts and forms introduced to me during my initial visits to the multinationals such as TQM, teamwork, customer satisfaction, productivity, cycle times, speed of execution, communication days and poster competitions, would keep emerging and re-emerging in one way or another during this time. Concurrently, notations of recurring themes and concerns connected to the interest in development, management, quality, productivity and so on were kept throughout the tracing process; and folded-in as part of subsequent dialogues with interviewees or as cues for further institutional pursuit. In this manner a picture was drawn of issues, debates, values, community-based concerns and policy or institutional responses surrounding this focus on quality, management and productivity.

This guided-tour through the terrain of official Malaysia that led me from the private confines of the two multinationals to numerous statutory agencies and so on, I have come to think of as "tracing the productivity and quality trail." While seemingly haphazard at first, I soon realized as the names of certain institutions and certain references (e.g., training, standards, efficiency) kept cropping up that this information given to me was a central part of the transfers and development nexus I was trying to document. Put another way, this mapping and tracing of the terrain became, in effect, a way of *executing* institutional ethnography in the field. In the process, a profile of institutions currently involved in generating, sustaining, regulating and/or deploying quality management know how in Malaysia was thus identified and approached.

MULTIVOCALITY: WRITING-IN MULTIPLICITY, LOCATION AND THE "I/EYE"

The textual and *re*-presentational (Radhakrishnan, 1996) equivalent of institutional ethnography, multivocality is not only a textual and located way of accounting for and writing-in the multiplicity of the terrain and phenomenon making-up our subjects and objects of inquiry, it is also a way to simultaneously problematize and account for silences attached to official stories. It does so by recognizing the *multiple* locations and intersections that individuals and institutions inhabit: for example, a manager who is both Malay and member of a profession, a researcher who is both observer and participant, a development strategy that is Universal in prescription but local in constitution. In a like manner, it does so by remaining equally attentive to *other* voices currently not accounted for in official stories; that is, those not directly engaged in an event or interaction, but whose worlds and lives may be or are organized by the visibilities and actions of individuals, institutions and discourses *not* immediately apparent in *their* world. *The old men on becas, schoolgirls walking home, and absence of cows so to speak.*

In the case of this document, the unofficial stories interwoven and inter-spliced within the official text mentioned previously [in chapters four, five and seven], and the active attempt to include as part of the story, an account of the research process and living experience or living history forming part of the project—whether connected to theory generation, fieldwork or writing itself—are particular instances of what I refer to here as multivocality. On a more conceptual level, this writing-in of the research process and researcher action is a central component associated with the recognition that both are important constitutors and mediators between so-called reality and our knowledge of this reality. It is a formulation underscored by, and linked to, as I note in my earlier discussion on institutional ethnography, the central role which people play in knowledge apprehension, knowledge use and knowledge creation activities.

Of particular importance here is attendance to the specific and multiple locations that people inhabit. Rooted in the recognition that apprehensions and representations forwarded by people come always mediated by education, by training, by class, by gender and/or by history for instance, this focus on location is attached to certain questions pertaining to the politics of representation. Include here: questions regarding who speaks, where such speaking is coming from, and who such speaking is geared towards. It is a way of acknowledging and taking accountability for the "politics of location" (Rich, 1984), and "situatedness of knowledge" (Haraway, 1988) highlighted by scholars in their critique of the social sciences and how it "knows." From the standpoint of institutional ethnography (Smith, 1987), it is part of the process needed to problematize and situate the everyday.

According to Radhakrishnan (1996, 1994), this acknowledgement and active confrontation over the relational and power-encoded nature of representers' location *with respect to* their subjects and objects of inquiry is central to any attempt at *re*-presentation. Somewhat akin to Kevin Dwyer's (1987) account of the dialogical nature of encounters between anthropologist and informant, encoded within such a constitution of the representational act is the recognition that knowledge is always produced *relationally* through a simultaneous process of negotiation and constitution between the writer-observer and subject or object of knowledge. To acknowledge and recognize this is to take accountability for the specificity and politics involved in any act of representation or observation. It is in short, a way of taking ownership over the social conditions giving rise to the production of knowledge (Radhakrishnan, 1994).

The political imperative galvanizing critiques such as the above and recognition over the mediatory role that knowledge constructors play in representing the Third World is particularly significant given the continued practice—by international and cross-cultural scholars of one kind or other—of relying on the existence and possibility of isolating *authentic* representations of Others who can be made known to the inquiring researcher from the West. Within these latter writings, culture and difference are often reduced and identified with national boundaries, where demarcations of a geopolitical nature are taken to be culturally *absolute, fixed and homogeneous* within (e.g., Boyacigiller and Adler, 1991; Schneider, 1989; Hofstede, 1987, 1985). As I indicate in my review of the literature, the implications attached to such a constitution on an international or national level are great—not the least of which is a reduction of what may well be diverse populations within national boundaries to standardized Others who can be wholly and fairly represented.

Categories of stratification and identities may well not be a fixed regime in how we constitute ourselves; and power or the "powerless" may likewise not be a fixed spectrum on a single axis. More significant however, is how a focus on the multiplicity of locations can aid in incorporating a variety of issues through precisely the recognition that time, race, voice and class may *all* be relevant in understanding any one situation, event and so on (e.g., Calás, 1992; Mohanty, 1991). As I have argued elsewhere (Chio, 1993), in societies as multi-cultural and/or diverse as Malaysia, where ethnic, cultural, economic, political and religious differences are constantly bisecting and intersecting one another, identities may have less to do with fixed signifiers like race, class, profession and nation, than they do with the ever *shifting* location of people's relations with one another, with the state, with religion and the like. There may well *not* be a definitive Malaysian [or Indian or Chinese and so on] that we can talk about, or to and or with.

This is significant since it is still widely believed and accepted by some that cross-cultural collaborations with local counterparts (e.g., Boyacigiller and Adler, 1991) can help avert the problem of non-understanding, and so confer on the research itself some modicum of authenticity: through in this instance, the supposed "insights" to the local that this local counterpart will provide to the research process and subject. The problem(s) attached to this line of thinking is the assumption that *all* locals are not only *alike,* but that there is an *endemic equality* in representation and voice that exists within a clearly locatable and observable territory like a nation.

According to Chatterjee (1993: 18), this feature and assumption is central to why modern nation states and their stated objectives of equality were "destined to fail." That is, the existence of the modern nation must necessarily—by virtue of its centralizing and bounding tendency—depend and fall back on a policy of "colonial difference" (Chatterjee, 1993: 16) in order to make possible the official representation and leading function connected to *being* a modern nation. This right-to-representation is determined, more often than not, by a whole slew of criteria and categories that locate what is widely accepted by modern society as authority, qualification, leadership and so on. It is in this sense that one can talk about location and voice as issues of central epistemological, political, social and economic importance.

Equally significant here in terms of how Others are known is the modernist dependence on certain objectification techniques, and the subsequent separation between subject and agent of knowledge. From the standpoint of knowledge and power, such procedures often mean that there is a strict representation of the Other, where the Self is absent from the representation (e.g., Escobar, 1994a; Prakash, 1990; Radhakrishnan, 1996; Said, 1979). It is a process and a practice that have led to the silent and invisible way by which "thought in relation to the Third World" (Escobar, 1992: 415) has been determined. As noted previously, knowledge constructors' reliance on a myriad of conceptual abstractions make possible not only their ability to know Others, it does so through a virtual negation of Others' existing diversity and living reality *while* obscuring the practices, concepts and relations which had already gone into the constitution of what is seen and what can be seen.

Note at this juncture that acknowledgement over location does not imply the absence of representation: what it does accomplish is to *acknowledge the act* of representation. As cautioned by Radhakrishnan (1994) in his article on postmodernism and the postcolonial project of historicizing and politicizing the present, there is a distinction and a reminder which need to be made here about metropolitan theory's strategy of self-reflection. According to him, the failure of grand narratives giving rise to postmodernism in the West, and their attending "epistemology of relativism" (: 309) is recuperated through "the strategy of self-reflexivity as a catch-all answer for cross-cultural crises or problems" (: 309). It is a reflexivity that, as he goes on to note, "*by itself does not* and *cannot* guarantee the *knowability* of other cultures and histories" (: 309, emphasis mine). This, to him, constitutes nothing more than another negation of the Other through knowledge: denying, in the process, the historical and relational manner by which Third World reality, and representations of Third World reality, had been constituted. It is a point similarly made by Calás (1992: 219) in her article on representing "Hispanic Woman": "making explicit the researcher's concerns . . . does not eliminate the rhetorical nature of the text," what it does is "incorporate explicitly as *rhetorics within the text* those elements that traditionally have been kept outside." As she goes on to note, the Other *still* cannot and *is not* speaking.

In sum, acknowledging the historical, relational and complicit connections between self and other, knowledge and power, producer and subject, is nothing more than a way of *bringing to visibility* this problem of non-accountability. It is equivalent, in a certain sense, to institutional ethnography's (Smith, 1987) intent of raising the specter of a problematic everyday *while still having to* remain accountable to the social conditions giving rise to, and produced by knowledge. However, by concentrating on the practices and relations that had and continues to produce and/or reproduce such conditions, this focus on intersections between the self and other, knowledge and power, producer and subject *do* bring to light the processes and the manner by which domination and its silences are produced and created. Multivocality is in short, a way to *begin* giving voice to the social organization of reality.

Chapter Three

Modernity and the International Management Gaze

> We need to anthropologize the West; show how exotic its constitution of reality has been; emphasize those domains most taken for granted as universal (this includes epistemology and economics); make them seem as historically peculiar as possible; show how their claims to truth are linked to social practices and have hence become effective forces in the social world.[1]

I turn in this chapter to the practices of knowing/knowledge construction by and through which Others are made known and knowable. As a key social organizer (Smith, 1987) framing and constituting current understandings and beliefs about knowledge transfers, I focus specifically on the international and cross-cultural management literature. Part of the institutional ethnography making up this project, the chapter is aimed at making visible some of the social particularities that define this literature as a way of knowing and doing, including its epistemic heritage and discursive entry-points. Thus, my focus here is on *how* this literature knows, as opposed to what it knows: that is, its content. In doing so, this review and analysis brings our attention to the problematic nature of the field's knowledge and claims, and the socially organized terrain constituting both field and knowledge.

I begin with a brief narrative of the connections between development studies and international management knowledge. This is followed by an introduction to recent deconstructions of development and postcolonial historiographies. Intended as a contrast to the preceding narrative, this juxtaposition highlights the problems connected to how modern development knowledge "knows" its subjects and objects of inquiry. The approaches taken by deconstructionists and postcolonial historians provide the subsequent framework for reassessing the writings and claims of international and cross-cultural management scholars. In reviewing this literature thus, the framework shows how it is that the Other of our knowledge can only be rendered visible and understandable in terms of a narrowly pre-specified coterie of concepts and causalities. Analysis of the literature from the standpoint of modernity make up the rest of the chapter, which ends by considering the implications of undertaking such analysis in terms of the methodology of institutional ethnography.

DEVELOPMENT AND INTERNATIONAL MANAGEMENT

At the center of modern development efforts is a focus on the overarching problem of Underdevelopment. Identified with a plethora of images and social conditions like poverty, illiteracy and disease, it is a condition most often associated with the Third World. As that corpus of knowledge and science devoted to unveiling the reasons for such underdevelopment, development studies is concerned with the search for effective antidotes to resolving this problem. Central to the claims of the field is a belief in the possibility of aiding Third World nations' *transition* from poverty-ridden and illiterate entities to developed societies able and capable of sustaining themselves. Development itself is a concept and a process involving the *planned evolution* or transition of these societies, whether from traditional/agricultural to industrialized, or from precapitalist to capitalist end-states (Lewis, 1978, 1966, 1955; Rostow, 1960).

Associated with these assessments over the conditions of, and solutions to underdevelopment has been the concerted and progressive formulation of an ever-growing plethora of development programs such as integrated rural development, poverty eradication, women in development and institutional development; each geared towards transforming specific aspects and/or social targets identified with the so-called problem of underdevelopment: in this instance, agriculture, education, women and institutions. Often paralleled by an interest in hard investments in infrastructure (e.g., Israel, 1987), such planning and program formulation relies on knowledge and capabilities *already in existence* within industrialized and developed societies, including science and technology, institutional or administrative knowledge, in order to help effect the transition of societies towards a developed end-state.

Within this backdrop, connections between the development and management sciences have been centrally organized by a focus on effectiveness and efficiency. This is particularly so in relation to the effective implementation of program or policy objectives, and efficient allocation and use of human and financial resources (e.g., Israel, 1987; Kiggundu, 1989). From the 1950s to the 1970s, this focus and concern translated into a reliance on modern administrative and management knowledge to affect within recipient countries a greater capacity for, and proficiency in implementing program objectives and resource utilization. According to Israel (1987: ix):

> . . . many of the real problems with development programs and projects lay not so much in their design and planning as in their implementation. Moreover, the real implementation problems were mostly institutional and managerial—in other words, rooted in the difficulties countries have in getting things done.

Mediated by this concern over implementation and the "effective use of" "human and financial resources" (Israel, 1987: 1), this reliance on and belief in administrative and organizational effectiveness became in itself a development focus throughout the 1950s and 1970s, articulated here through public administration institutes established by the Rockefeller and Ford Foundations, and institution building activities of the U.S. Agency for International Development, among others.

During this time, the procedures and practices for managing and organizing resources, which are so central to the claims of the management sciences as that discipline most vested with knowledge on how to do things efficiently and effectively (e.g., Barnard, 1938; Fayol, 1949; Taylor, 1911), offered to development studies the instrumental *and* conceptual entry-points for planning and developing the Third World. It is a connection and an entry-point that remains to this day. As exemplified by the writings of Moses Kiggundu (1989) and others (e.g., Esman, 1991; McBeath, 1990), the critical operating and strategic management tasks informing the activities of modern organizations are crucial for the development and modernization of developing country institutions and economies. As Kiggundu notes in his assessment on why this connection between development and management exists:

> . . . the problems of *implementation* and *sustainability* of development initiatives are pervasive and persistent. . . . [raising] . . . concerns [about] organization and management for economic development and social change. . . . Experience over the last three decades has clearly shown that the constraints to development have little to do with overall availability of resources. Rather, the most serious constraints relate to *the organization, management, and utilization* of available resources and the opportunities for more balanced development (1989: xvi—xvii, emphasis mine).

As with the analysis offered by Israel (1987), the problem of development is not so much a question about the efficacy of modernization or development knowledge, be it concept or model. Rather, it is a question of *not* having done it right. Indeed, the importance of administrative and organizational knowledge to rectify the problems of the Third World is so great that failures of the "tried and true" recipes of development, if acknowledged at all, are reduced to a(n) *improper, insufficient,* and *mishandled* implementation of development projects (e.g., Amin, 1990; Escobar, 1994b, 1988, 1984; Manzo, 1991).

In more recent times, this connection between development and the management sciences has been extended and organized by an explicit belief in markets, and the "transformative power" (Berthoud, 1992: 79) of markets to effect the transition of developing nations into developed entities. As noted by the *World Development Report* (1991), countries failing to improve their economic performance do so because they have relied on means of resource allocation other than mechanisms of the market.

Triggered by the dismal management of macroeconomic policies and growing disenchantment with public or state-owned enterprises whose performance had generally fallen below expectations—represented here by poor economic results, overstaffing, high debt ratios, and/or the non-responsiveness of top management (Edwards, 1995; Israel, 1987; Kiggundu, 1989; Lieberman, 1990]—this market-mediated focus has been at the center of structural and sector adjustment policies (SAPs) of the World Bank since the mid-1980s. The economic success of certain East Asian nations (e.g., Korea, Taiwan, Singapore, Hong Kong) throughout the mid-1980s to the early 1990s only seemed to confirm the efficacy of what development analysts (e.g., Edwards, 1995; Lieberman, 1990) and the World Bank (1993a: 9) call a "market-friendly approach" to development.

Thus we see at the center of adjustment policies, a concerted effort to restructure Third World nations into economic facsimiles of the industrialized West—open to and supportive of international trade (Edwards, 1995; Kiggundu, 1989; Lieberman, 1990; World Bank, 1991). As a planned transition, SAPs focus on the "comprehensive restructuring, management, and coordination" of developing country institutions (public and private, social and economic), which entail the "gradual but steady" "elimination of market imperfections" while allowing "competitive market forces a bigger influence in the making of economic decisions and allocation of resources to organizations" (Kiggundu, 1989: 257). To this end, policies encouraging deregulation and/or privatization are central components of SAPs, as are the acquisition of modern management and organizational knowledge and practices.

Like their development counterparts, management scholars focusing on this connection between markets and development (e.g., Esman, 1991; Frankenstein, 1986; Jorgensen, Hafsi and Kigundu, 1986; Kiggundu, 1989; Segura, 1988; Sood and Kohli, 1985) see the systematic privatization and rationalization of previously non-market and/or developing economies as *the* antidote to poverty, illiteracy and economic backwardness. In their views, *privatization inculcates efficiency* through the *combined* forces of the market (e.g., Henley and Nyaw, 1986; Jorgensen, Hafsi and Kiggundu, 1986), and the management technologies such an organization of the economy implies (e.g., Kiggundu, 1989).

Under this more recent constitution of the problem of development, the management sciences, and its location as that genre of expertise specializing in designing, producing and sustaining efficient institutions in market economies, is explicitly linked to development in form *and* strategy (e.g., Bremer, 1984; Dror, 1984; Esman, 1991; Jorgensen, Hafsi and Kiggundu, 1986; Kiggundu, 1989, 1986). In other words, management knowledge is no longer just that compendium of procedures that will allow for the more effective implementation of program objectives or efficient use of resources; it is now *the* target and form of modern knowledge and practice needed to transform the Third World. Hence, the focus on, and interest in transfers of management expertise making up the entry-point for my project.

In sum, connections between development studies and the management sciences have been constituted by a mutuality in terms of what both disciplines take as central and instrumental for effecting the development of Third World nations. At the center of this connection is a prognosis linking management knowledge, efficiency and the development of nations. Armed by the respective claims of their fields about the planning capacity of development experts, and the effectiveness and efficiency of management know how, the prognosis offered confronts how Third World organizations *and nations* can be better managed to ensure their transition towards a developed end-state, through, in this case, SAPs and privatization. Informing this capacity for change is a belief in the transformative power of modern knowledge (e.g., science, technology, management), and social structures (e.g., markets). The possibility for planning itself provides to development and management experts their conceptual and instrumental entry-points for engaging in the terrain of Third World nations.

DECONSTRUCTING DEVELOPMENT: THE POLITICS OF REPRESENTATION AND THE THIRD WORLD

Unlike the prescriptive story on how developing countries can be actively planned and restructured out of their underdeveloped conditions noted above, recent deconstructions of the development sciences highlight the debilitating effects that such a form of knowing have had on the Third World (e.g., Crush, 1995; Escobar, 1994b, 1988, 1984; Ferguson, 1990; Mueller,

1987a; Sachs, 1992). According to these deconstructions, the Underdevelopment of the Third World which discourses of development claim to know and want to solve is first and foremost a creation of the discipline, and of those institutions—epistemic, scientific, professional, economic—up-holding it as a Modern field of knowledge.

Rather than accepting underdevelopment as a natural and pre-existing condition characteristic of Third World societies, these deconstructions focus on the social conditions of knowledge construction, and how these conditions have *led to* the production of a problematically underdeveloped subject who then becomes the object of attention and intervention. For these writers, representations produced by development discourses are effective social and cultural organizers of sight and action, relying on very specific constitutions of reality in order to do so.

This power to define reality and sight is doubly important in the context of the Third World however because modernity is not only about ruling practices (Smith, 1987), it is also about the constitution of reality in a very concrete and practical sense. As noted by Escobar (1994b), there is a "materiality" that is attached to the ways the Third World has been and still is constituted, economically, institutionally and discursively. Whether culled under the rubric of discursive practices, institutional and professional conventions or procedural activities [and interests] of western institutions of one kind or other (e.g., development, multinational), the social terrain of Third World societies are configured and re-configured, transformed and reformed, through the concrete practices of a plethora of experts whose everyday activities—of problem identification, assessment, planning and decision-making—are key conditions of possibility leading to the material production and re-organization of this terrain. Put another way, the focus on knowledge construction and constitution of reality is not just a matter of conceptual apprehension, it is centrally connected to practice, and results in the elaboration of yet other forms of practice whose effect is the reconfiguration of life, labor and language (Escobar, 1994a).

Moreover, there is an historical relationality attached to such representational and material procedures. This has to be acknowledged since much of what is understood and identified as Third World, for example, poor, illiterate and/or backward, is often tied-up and bounded by the legacies of European colonialism and modern imperialisms of one kind or another. As noted by Radhakrishnan (1994), the history of the postcolonial and Third World Other is marked in *fundamental relation* to that of the West. Contrary to the linear and progress-oriented character of modern narratives, this story about self and other, development and underdevelopment, is one fraught with disparity and paradox.[2] From the writings of political economists and dependency theorists to more recent Third World feminist, development or postcolonial critiques, scholars highlight the debilitating effects, intentions and silences—whether epistemic, scientific, cultural, economic and/or political—locating relations between the West and the Rest.[3]

These critiques focus our attention to the manner by which Third World societies have been systematically rationalized by the knowledge and policies of the West, setting into play a vast network of relations: notably, between institutions of knowledge construction and the Third World, between development agencies and Third World nations, between national governments and its citizenry, and between western capital and non-western labor. The net effects of such relations, including debt crisis, invisibility of local knowledge, marginalization of labor, dislocations of culture and communities, constitute the terrain from wherein these societies must subsequently negotiate with the West and amongst themselves.

As an example, development policies tailored towards the mechanization of agricultural activities in Egypt by the U. S. Agency for International Development (Agricultural Mechanization Project, 1979–1987) not only led to large-scale debt and dependency for the nation without much by way of increased yields, it also effectively de-politicized and neutralized larger issues of social inequality and powerlessness, both local and global, including in this case, a disproportionate access to and control over land by a few, and a compression of local wages and labor. According to Mitchell (1995), there was a politics and a political economy organizing these policies whose aims it was to promote the transition to open markets, and the purchase of equipment from US companies. Procedurally, such an intervention was made possible by a construction of Egypt which textually and discursively framed the nation as an entity de-linked from either history or a social location within the global arena, and objectified it as a "map-like" entity composed of and caught between the dual problems of overpopulation and limited arable land.

Noted somewhat differently by Chinua Achebe (1976) in his critique of the colonial critic, one of the central conditions of possibility enabling European and Modern domination over the colonized is the progressive ability and assertion by colonizers that they *can* and *do* know the colonial Other. The development expert assessing and acting on the Third World is thus only the most recent variant on this relation and this history, one whose authority is based on his or her claims to knowledge. These claims to a privileged ability to know and represent the Other to themselves is ultimately an exercise of power and domination which defines and subjugates the colonized through its ability to *control* and *set the parameters* of the subsequent "conversation" between colonizer and colonized.

Indeed, this power to set the "terms of engagement" is a historical relation that has been a central focus of the subaltern project (e.g., Chatterjee, 1993, 1988, 1987; Chakrabarty, 1992; Pandey, 1990; Guha, 1983) to contest dominant constructions about the Third World as passive recipients of history. According to these authors, subaltern historiography is less a voice from the Other than it is a *response* by the Other to constructions about them. Indeed, subaltern historicity is forged in relation to, and within, the parameters of a history that is singularly European in its dependence on certain modernizing narratives (e.g., nation-state, science, progress), and a position of subalternity and dominance which is built-into and can only be articulated in the name of this same history (Chakrabarty, 1992).

From this standpoint of representation and voice, institution and practice, history and relations, Modernity is primarily a political project of knowledge and power. To recognize this is to know that there is a politics of representation making-up and making possible modern knowledge and their claims to know. According to Escobar (1993, 1988, 1984), understanding modern discourses in terms of this politics of representation entails attending to the forms of knowledge that refer to them, the system of power which organizes their practice, and the subjectivities they foster.

In effect, Development as a modern order of representation is organized around a constellation of concepts such as poverty, production, growth, planning, market, science and technology (Escobar, 1994a, 1988, 1984; Sachs, 1992) without whose existence the very notion and practice of development as it is currently understood can have neither meaning nor visibility. Informing and organizing these concepts are very specific forms of knowledge, practices, values and assumptions that up-hold and produce the discourse, its claims to knowledge, its prescriptions, and its representations of underdevelopment and the underdeveloped in the Third World. In the words of institutional ethnography, to understand the problematic nature of development studies and the management sciences is to refer to Modernity and the intersections between culture, knowledge and practice which organize these discourses and the representations produced and deployed by them.

Modernity and the Intersections between Culture, Knowledge and Practice

Coming into focus in eighteenth century Europe, Modernity refers to that Enlightened era when "[m]an [sic] developed forms of knowledge about himself and about the world" that required a "distancing, objectifying posture" (Escobar, 1992: 67). Epistemologically, it is marked by the birth of reason and particular forms of rationality. Culturally, it entailed the separation of life into distinct spheres like the economy, culture and politics, with each the owner of a science devoted to revealing the secrets of society. Economically, it was built on the consolidation of capitalism. And politically, it signified the advent of democracy.

As a "scopic regime" (Jay, 1988) and a "scientific gaze" (Escobar, 1994b), Modernity elevates to ontological status Western ideals about progress, reason and science. Also encoded in modern ways of seeing and knowing are certain beliefs and assumptions about time (evolutionary, lineal), causality (functional, objective), value (economic, productive), space (identifiable, concrete), gender (male/female) and so on.

We find as an example, in referring back to the features and prescriptions of the development sciences, that the field's constitution of development as a process entailing the *transition* of nations from an ostensibly undeveloped to a developed end-state, a product and a reflection of the modern belief in evolution and progress. Concurrently, its constitution and evaluation of problems and solutions *in terms of* efficiency and effectiveness is a feature traceable to modern notions about value and economy, as is the focus on resources, resource utilization and the like. Finally, the focus on planning and the planned approach to development—logical, systematic, step-by-step—which underwrite the claims and prescriptions of development knowledge, is already framed and en-framed by modern beliefs about logic and rationality.

In this sense, modern knowledge such as development studies and the management sciences are not only products of Modernity's gaze, they constitute an embodiment of how and what it sees itself as representing (e.g., Escobar, 1994a, 1991; Harvey, 1989; Mitchell, 1995; Radhakrishnan, 1994). Thus we find that development scholars' beliefs in evolution and efficiency are what make possible their understanding and ability to see or categorize societies as economically developed or undeveloped; their privileging of planning, indeed, their capacity to plan an outcome of the rationalist and scientific claims of the field. It is in this way that modern know how embodies what and how their knowers represent. Knowledge so understood is a *cultured* visibility.

Modern knowledge, however, is not just a cultured knowing; it is also a rule-governed *practice* (e.g., Escobar, 1994b, 1991). It is composed of a process involving activities like problem identification, analysis and sense making; and constituted by acts/activities—such as data gathering and analysis—that rely on very particular techniques in order to know. Techniques of application or deployment include the demarcation of disciplinary boundaries, the persistent separation of reality into distinct social spheres, the specification of relations, the objectification of reality, dependence on scientific measures and so forth (Escobar, 1991). It also relies on a separation between the object and subject of knowledge, and a dependence on binary oppositions like developed-underdeveloped, self-other, efficient-non-efficient, in order to produce its knowledge claims, whether in the form of descriptions, causalities or prescriptions. On an assumptive level, it depends on and replicates those privileged beliefs noted about time, economics, knowledge and so on noted previously.

In the particular instance of the development and management sciences, also included as a technique of rationalization is an *economism* that is silently locating our apprehension and representations of reality and of Others (Escobar, 1994b; Touraine, 1995). As noted by Escobar (1994b), development as a rule-governed gaze not only rendered farmers, women and nature visible to the development apparatus through its discourses and programs on Integrated Rural Development, Women in Development, and Sustainable Development, it also dictated *the manner* in which these "problems" (i.e., farmers, women, nature) were to be apprehended: namely, as economized factors and/or problems to be solved. According to him, the economized constitution of reality that is in the development sciences ensures in this instance that farmers, women and nature are en-framed and packaged in a manner which would make their "consumption" by the development apparatus understandable, and so "digestible." In the process, the living complexity and history of farmers, women and nature are silenced and reconstituted into narrow and pre-specified terms.

Central to this possibility of silencing those that development *sees* is the modernist practice of rendering the object(s) of knowledge *external* to both its representation and the knower self (e.g., Burrell and Morgan, 1989; Escobar, 1994b, 1991; Mitchell, 1995; Smith, 1987). In particular, the reliance on pre-specified categories and forms of analysis—economized, scientific and so on—silences the cultured and constructed nature of how we know. These analytical forms and categories are accepted as *real* when in fact they are no more than Western explanations about Others to the West (e.g., Prakash, 1990; Said, 1989, 1979). In short, they reflect what we as knowers see as important, but come already constituted by our ways of knowing (Smith, 1990, 1987).

Noted earlier, this attendance to knowledge and its social conditions of production is a central focus of postcolonial (e.g., Achebe, 1976; Chakrabarty, 1992; Chatterjee, 1993, 1987; Pandey, 1990; Prakash, 1990; Radhakrishnan, 1996, 1994), and development critiques (e.g., Esteva, 1987; Nandy, 1989, 1988; Parajuli, 1991; Rahnema, 1988a, 1988b; Sheth, 1987; Shiva, 1990, 1989, 1987; Visvanathan, 1987). According to these authors, the concepts and blind transposition of singularly western and modern categories of analysis have served to standardize and normalize Third World experiences and differences, whether in culture, identity, experience, history and so on. In particular, knowledge construction practices and the techniques of application they rely on such as binary oppositions and scientific or economic reductions, have objectified and rendered to the gaze of this knowing West, a Third World Other who is, and can only be known according to these terms.

It is to this simultaneous procedure of [re]-presentation and silence, inclusion and exclusion, that Mitchell (1988: 32) refers when he observes that a consequence of this "regime of objectivism" is the subsequent division of the world into two realms, "a realm of representations . . . [and] exhibition," comprised of the Other as the object and subject of modern inquiries, and "a realm of the "real'" in which resides "external reality" and the modern Self as inquirer and knowledge constructor. In the process, we accept as Truth our representations of these others, and proceed to develop systematic prescriptions on, about and for them.

This, however, is still only part of the story. For not only is knowledge construction a rule-governed process, it is just as systematic in the manner by which its claims are deployed (Escobar, 1994b, 1988, 1984). This regulated and rationalized constitution and proliferation of knowledge through standardized concepts, causalities, problem identifications and practices govern *fields of knowledge* such as the development or management sciences, and *fields of practice* like consulting, management and academia by ensuring how subsequent problems, issues and interactions are to be apprehended, explained, organized, produced and so on. In sum, visibilities are produced and reproduced through this inscription of knowledge and practices into institutionalized conventions, and acted-on or redeployed through a corpus of institutions such as the World Bank, the International Monetary Fund, local development agencies, and academies of sciences, and *their* further in-site activities (e.g., Escobar, 1994b, 1988, 1984; Ferguson, 1990; Mueller, 1987a).

This overview of development deconstructions and postcolonial historiographies provides an alternative constitution and understanding about the "methodologies" and practices that underwrite modern discourses. From the standpoint of institutional ethnography, the insights derived from such a *situated* analysis are central to the aim of problematizing and making apparent those relations and practices which are constantly organizing phenomenon we take for granted, and our knowledge of it. In particular, the focus on culture, knowledge and practice provides an important starting point for knowing what to attend to in our attempts to decompose how and what we know.

THE INTERNATIONAL MANAGEMENT GAZE AND TRANSFERS

The frame of analysis and critiques previously discussed is deployed here over the discourse of transfers in the field of international management. Under the auspices of this frame, we see that preoccupation with transfers and the Other is not just an exposition on the "what," the "why" and the "how-to" of development. Rather, attention is pinpointed to *how* this story of transfers as represented in international or cross-cultural management literature is made possible, and so, *tell-able*. Within this story, what is visible is a series of cultured and modern assumptions, and a certain approach to knowledge and to objects of inquiry—functional, objectivist, reductive—that is the core of the story as told from this standpoint of knowledge, culture and practice.

The Construction of Subjects

As is the case with its development counterpart, the active construction of an Underdeveloped Third World subject is central to the discourse on transfers of management knowledge and expertise. This construction is made possible not only through the presence of certain assumptions and values, but also by certain concepts around which the field of management studies is organized. These concepts *mediate* the specific visibilities of the management scholars in their construction of the Third World subject. From the standpoint of mainstream international management literature, knowledge transfer as a solution to underdevelopment is basically organized around concerns regarding the formation of institutions, the efficient management of institutions, and the transfer of the managerial or organizational expertise needed to manage these institutions. This focus has been, and is, a central premise underlying current models of development.

Within this story, there are a series of prescriptive norms. Transfers of knowledge are seen as a necessary response to certain problems of "absences" and "inefficiencies" (e.g., Esman, 1991; Israel, 1987; Kiggundu, 1989; McBeath, 1990). In particular, the *absence* of modern administrative or management expertise (e.g., Erden, 1988; Jones, 1989; Jorgensen, Hafsi and Kiggundu, 1986; Khandwalla, 1989; Kiggundu, 1989; Lo, Yau and Li, 1986; Molz, 1990; Psacharopoulos, 1984; Safavi, 1981; Segura, 1988; Warner, 1986; Yavas and Cavusgil, 1989), and/or the *inefficiencies* attached to *existing* organizational forms within these developing nations (e.g., Khandwalla, 1989; Kiggundu, 1989; Kumar and Singh, 1978; Sood and Kohli, 1985; Vernon-Wortzel and Wortzel, 1988).

Nowhere is this problem identification as noteworthy as Kiggundu's description of the *"typical"* organization in developing countries:

> As *anybody* who has visited the executive suites of these organizations knows, top management tends to be overworked . . . [a condition connected to] the reluctance to delegate. . . . Administrative and technical support is weak. . . . Quite often the organization's mission is not clearly articulated and the organization lacks a clear sense of purpose and direction. . . . [F]or most of these organizations the middle is weak, with weak management and organization systems and controls. There is little motivation for independent action . . . because of middle management's lack of technical and managerial skills and experience. In return, [they] exercise close supervision [as do their own bosses]. . . . The operating levels are inefficient and costly . . . with overstaffing . . . [and employee] underutiliz[ation]. . . . Pay is poor . . . [leading to] . . . [poor] morale . . . [which] manifests itself in unexcused absences and worker indolence . . . (1989: 9, emphasis mine).

Through such identifications, descriptions of Others render visible and tell us something about them *as* recipients of transfers: namely, their lack of management know how and inefficiency. According to the above, Others are characterized by a very specific coterie of organizational problems like weak administrative and technical support, inefficient and costly systems and controls, and overstaffing. Also included are problems of a managerial nature such as their lack of direction, and reluctance to delegate. Finally, problems identified are also human in scope, specifically, local workers' lack of motivation, indolence and poor morale. Concurrently, these identifications also forward to us a message on what is important or crucial to have, either organizationally or otherwise: for example, delegation, strong administrative and technical support, a clear sense of purpose, motivation, efficiency and cost effectiveness.

Within the identifications and stories told here, Underdevelopment is signified by whether Third World subjects either *have* or *do not have* these features, processes and/or knowledge that their authors consider important, whether in terms of the disciplinary field or the culture they come from. Values and concepts like efficiency, delegation and cost effectiveness are filters and mediators *by* and *through* which these authors identify and capture the nature of Others' underdevelopment. In the process, not only do these normative assessments provide a way of calibrating and categorizing Others, they also make possible our ability to see Others *as* the *objects of,* and *reasons for* transfers. In effect, Others become a location or a site where problems have to be solved.

From the standpoint of the analytical framework deployed previously, this identification of absences and lacks is a central feature of *how* subjects of development are produced and constituted as problems. Procedurally, their constitution relies, as I note above, on certain concepts and values that are central to the field in order to make possible this identification of Others. As articulated in Kiggundu's description, these concepts and values are key to his [re]presentation of, indeed his *ability to* [re]present to us, the underdevelopment of the Third World.

Other assessments and prescriptions provided by several authors linking management and organizational capacity to development include "better *coordi[nation]* of their [Third World] policies toward direct investment" (Weigel, 1988: 5, emphasis mine); and "ways of making existing *organizations* in developing countries *more effective*" (Jorgensen, Hafsi, and Kiggundu, 1986: 417, emphasis mine). In the former, the Third World subject is defined through the concept of coordination, in the latter, through the concept of organizational effectiveness. Concurrently, the reliance on and use of, normative and comparative words *in* these statements such as "are required" and "more effective" serve as constant reminders that coordination and organizational effectiveness *do not yet exist;* hence the status of these subjects as objects for development.

When Others are seen through such situated lenses, what is seen is less an Other than it is their location as a *site* for delegation, coordination, effectiveness and the like. It is in this sense that I made the earlier comment regarding how Third World nations' living reality *disappears* when seen through such concepts and values which are central to the observer's own lens.

Presented as simple descriptions or prescriptions, the presence of these values and assumptions are often *not* immediately apparent in the constructions of Others, even when they are *precisely* what is anchoring these constructions. This process of constitution and assessment under the guise of description is well exemplified by Geert Hofstede's account of the differences between cultures with an individualist (Western Self) and a collectivist (Mexican Other) orientation:

> A group of Mexican immigrants is employed by a US shoe factory. The Mexicans are hard workers but they (1) stick together, and individuals refuse to be transferred to other departments if the need occurs; (2) do not speak up in meetings with others present; (3) are suddenly absent for fulfilling family obligations and religious celebrations not understandable to their management; (4) however, what management likes is that they don't want to join the US workers' union (1985: 355).

Encoded in this presentation is an opposition between Self and Other, and an implicit judgment about the Other which leaves something to be desired. The image that emerges in a reader's mind upon reading this description is a constitution of the Other as collusive [*they stick together*]; non-cooperative, non-participative or non-assertive [*they refuse to be transferred when needed, they do not speak up in meetings*], and irresponsible [*they are suddenly absent*]. The reason why this is so is because the author and reader, or intended audience of primarily western management scholars, researchers and students, *already accept* as important and un-problematically normal, certain values and assumptions which are silently locating the "description." Included in these are the positive values given to individualism, independence, cooperation, participation, vocal proficiency or expressiveness, responsibility or dependability, ability to separate the public from the private and so on. While Hofstede's analysis may have been presented or intended as a neutral description of Self and Other differences, *in the process* of undertaking this presentation, it ends up reinforcing an image of the Other that is negative, and the Self that is positive.

From the standpoint of institutional ethnography, this look at the cultural and social "surround" that are constantly framing and organizing our apprehension and representations of Others serve as an important reminder about the located nature of what and how we know. It is a reminder that knowledge is always coming *from* an elsewhere and a somewhere that may not be immediately apparent to us at the moment, indeed, even during the moment that we are looking and/or describing.

The silent and value-coded judgments I bring attention to here is a key feature of the objectification procedures enabling modern forms of knowledge to not just describe, but to produce the objects and subjects of their inquiry (Escobar, 1994b; Prakash, 1990). It is a process reinforced by the reliance on binary oppositions whose effect is the equally silent relegation of Others to positions of inferiority, and the Self to positions of superiority, however implied.

Similar to Hofstede's (1985) descriptive-valuation above, the following example illustrates how this marginalization of Others are made possible because certain values are already silently *anchoring* the [re]-presentations and oppositions that authors rely on to articulate knowledge. In his call for China to transition towards a market-centered economy, Frankenstein (1986: 148) describes existing structures in the nation as "politicized and secretive, bureaucratic, and consensus-driven." The specific descriptors used here to locate China are culturally coded against it because of the implicit negativity already associated with these descriptors. This is particularly so when juxtaposed *in relationship to* the "profit-seeking free market business practices" (: 148) of an industrialized Self. This is a positive constitution made possible by relying on the *association* of these descriptors (profit-seeking, free market) to values about openness, rationality, organizational flexibility, individuality and independence so central to modern notions of self and society in the West.

This silent and hierarchical assessment recalls the evolutionary assumptions in the writings of certain modernization theorists in development studies. For instance, W. W. Rostow (1960) wrote about "stages of growth," which involved the transition of nations from traditional backward to advanced industrialized states, and W. A. Lewis (1978, 1966, 1955) the "economic organization of production" involving the evolution of nations from precapitalist-agricultural to capitalist-industrial states. Writings such as Frankenstein's (1986) and others noted previously (e.g., Esman, 1991; McBeath, 1990; Kiggundu, 1989; Israel, 1987) focus on, and are already embedded in a like belief about evolution, albeit mediated here by the focus on management know how and market rationality as *the* mechanisms for transforming nations from undeveloped to develop end-states. In this case, developed end-state refers to a very particular organization of society and social life that is industrialized, free-market and capitalist. Thus, Frankenstein's (1986) opposition between Self and Other is already mediated by the author's belief in evolution *and* markets as ways of understanding and structuring society.

Finally, the explicit and implicit economism inherent in the discipline's concepts and values is another objectification procedure by which Third World subjectivities are constituted. An endemic feature of the discourse of the management sciences, it reflects the discipline's location as a genre of expertise linked to modern capital. Referring back to descriptions discussed above, we see that descriptors and categories used by authors are already economized through concepts such as resource utilization and investments. In this regard, their interests in efficiency, costliness, profits and market forces are likewise economized.

Accordingly, the literature relies on an almost wholesale treatment of both its subject and object of knowledge in economistic terms. In the following example, the value of developing nations appears to exist only in their location as an inventory of assets and human resources:

> Developing countries have three great assets: the land, the gods, and the people. . . . [P]rospects for social, economic, and political developments for most developing countries are intimately related to their ability to develop and effectively utilize their human resources (Kiggundu, 1989: 143).

It follows that most countries outside the industrialized bloc are objects of interest primarily because of their location as resources or cheap labor (e.g., Kiggundu, 1989), or more recently, as potential markets (e.g., Lo, Yau and Li, 1986; Vernon-Wortzel and Wortzel, 1988, 1987), partners in joint ventures (e.g., Christelow, 1987; O'Reilly, 1988), and/or competitors (e.g., Grant, 1991; Portcr, 1990, 1986; Thurow, 1992).

Nowhere is this economism more real than in recent globalization literature (e.g., Bartlett and Ghoshal, 1989, 1988; Hamel and Pralahad, 1989; Ohmae, 1993, 1990, 1989a,1989b; Osborn and Hagedoorn, 1997). To begin with, the very interest in globalization is one organized around explicitly economic concerns such as competitive advantages (e.g., Hout, Porter and Rudden, 1982; Porter, 1990), market participation (e.g., Yip, 1995), transaction costs (e.g., Buckley and Casson, 1988; Duning, 1993, 1989; Nooteboom, Berger and Noordehaven, 1997; Park and Ungson, 1997; Parkhe, 1993, 1991), and potential or expected returns associated with the innovation, research and development of products or information processing and technological capabilities attached to globalization (e.g., Goes and Park, 1997; Osborn and Hagedoorn, 1997; Teece, 1989). This connection between economics and globalization is not surprising since much of international and cross-cultural management's earlier interests in the potential of globalization were framed with a focus on economic opportunities (e.g., Henley and Nyaw, 1986; Jorgensen, Hafsi and Kiggundu, 1986; Kiggundu, 1989; Molz, 1990; Vernon-Wortzel and Wortzel, 1988), or economic limitations (e.g., Emmerij, 1992; Porter, 1990, 1985; Thurow, 1992).

Indeed, according to the following assessment, the economic imperative attached to globalization is central for *effecting* the transition of newly developing or emerging nations toward more developed and democratic end-states:

> Globalization has largely been due to worldwide economic development and the opening of domestic markets to foreign firms. In fact, economic change often leads political change. Economic development that *creates needs and desires* for business products forces politicians to agree on new rules to encourage further economic development and growth. . . . Currently, economic concerns *are driving* major political changes in Eastern Europe, Russia and China. The end of the cold war has opened Eastern Europe and Russia to new economies and developing market places. Additionally, China has been targeted for new investment capital by [international] businesses. . . . [such development and change] *make it easier for firms to enter* international markets. . . . Moving into new markets provides new opportunities . . . (Hitt, Keats and DeMarie, 1998: 23–24, emphasis mine).

Noted earlier, underlying such analysis is the economic logic galvanizing interest in neoliberal models of development. What is significant from the standpoint of the *economizing* features of the discourse and its constitution of reality is how this argument is premised singularly on a constitution of Others as modern market citizens. That is, Others remain *invisible* unless they can be located as *consumers* who are the "driving force" behind needs and desires for business products.

This observation about what is *missing* or *non-apparent* in descriptions like the above is a reminder that the procedure of economism I call attention to is also informed by certain social and economic particularities—for instance, the interests of capital, the location of the field in connection to capital, and the policies of development agencies—that frame and organize the claims of management scholars. This re-situated perspective is a reminder that part of what is at stake in concerns over globalization is an expectation of *remaking* societies into economic facsimiles of the industrialized West: in particular, the interest in reconstituting them into free-markets which will provide to existing businesses and industries *in* the West, new and potential markets for the flow of their goods and services. To this end, discourses of globalization themselves *do not just* document the phenomenon of globalization, they also provide an important conceptual, discursive and policy-oriented *entry-point* for opening up and easing the way for the interests I note here.

In the process, Others are *recast* in role and form into subjects whose value is derived singularly from their connection to modern organizations, and their participation in the modern economy. As indicated by Kiggundu's (1989) earlier description, it is in their effective roles as manager and worker that Others are the subjects of interest. Additionally, there is a very specific configuration associated with manager and worker which is of interest:

> . . . the lack of human resources constitutes the major constraint on the development of the region [Africa]. . . . The lack of skills and experience to man [sic] the modern public and private sectors has always been regarded as the major constraint on economic development . . . (Davies, 1980: 55).

It is the skills and experience of these individuals that is important; their possession or *non*-possession of these, a barometer of their *value* as human resources. To this end, identifications or assessments of Others in terms like motivation, job involvement and commitment (e.g., Putti, Aryee and Tan, 1989), work-related values (e.g., Hofstede, 1987), managerial job attitudes (e.g., Kanungo and Wright, 1983), management style or roles (e.g., Tayeb, 1987; Torbion, 1985), human resources (Ricks,

Toyne and Martinez, 1990), and so on firmly *locate* them as *labor, producers* and *organizational members.* Through the authors' focus on these values and features, these assessments provide barometers for specifying what is *desirable to have* in modern workers and managers.

In sum, not only is the discipline's interest in Third World Others explicitly economic in nature, so too are its categories of description, range of interests, evaluation criteria and constitutions of subjects. It is important to note here that these features of the discourse are also to be found in the context of the authors' constitutions of people and realities "at home." My argument is *precisely* about the implied knowability and applicability of primarily self-referential and self-interested categories to any and all contexts.

The picture that results from such constructions and non-reflective application of cultured and economized categories is the very "facticity" of the world as the modern West sees it. No matter *where* we look, it's about organizations and management and efficiency and markets and so on: a cartography of Others both modern and economic. As noted by Hirst and Thompson (1994: 288) in their critique of recent globalization literature and its preoccupation with one single market, and one single economy: "[t]he vision is of one large interlinked network of producers and consumers plugged into an efficiently operating "level playing field" of the open, international and globalized economy."

More important to note, however, is how this economization is able to effectively silence the living reality of those we study. As the following example shows, people are not only constructed as problems to be solved, but their very reality and complexity are *reduced* to economized and organizational terms:

> The major sources of low motivation reported by Zambian workers are (1) personal problems such as domestic quarrels, . . . alcohol consumption, hunger from inadequate food intake, and death or sickness . . . ; (2) unfair organizational practices; (3) poor relations with superiors, coworkers, or subordinates; (4) lack of fit between the job and the worker, . . . (5) inadequate intrinsic or extrinsic sources of motivation, . . . [and] (6) lack of opportunity for growth and advancement [(Machungwa and Schmitt (1983), as quoted by Kiggundu (1989: 169)].

In this instance, the living reality of Zambians, whatever this may entail, is already organized by the singular focus on them *as* workers. This constitution of Zambians makes understandable the subsequent discussion of their reality *only* in terms of organization and job-related interests in motivation, supervisory relations and so on. While an implicit recognition exists that they are also constituted by potentially *non-work* locations or concerns, instanced here by the reference to their domestic and social surround, even these ostensibly more social and political issues are reduced to an organizational and work-related level. In this case, it is a constitution made possible by the conceptual and documentary *reduction* of these social particularities to the status of "personal problems" directly linked to the low motivation of Zambian workers.

These conceptual mediations *hide* the culturally and socially constructed nature of the picture presented by the authors. The dependence on certain concepts presents to us very particular ways of knowing Others and makes possible the belief that Others *can be known.* In the process, these procedures and assumption render digestible the complexity and particularity of Otherness *into* the knowledge creation apparatus (Escobar, 1994b). In the example above, the constitution of Zambians by way of economized and managerial categories like "motivation" and "worker" make it possible for management scholars to *include* Zambians' hunger from lack of food intake in the same conceptual space as lack of fit between the job and the worker.

In sum, within the international management literature, the constitution of a backward Other is rendered possible through writers' dependence on concepts, practices and forms of knowledge which encode very particular assumptions and/or values traceable to the field's modern heritage (e.g., Argyris, 1957; Barnard, 1938; Herzberg, 1966; Katz and Kahn, 1966; March and Simon, 1958; Taylor, 1911). What gets put into play is a hierarchy, such as developed-underdeveloped, in accordance with these values that members of the discipline often use as starting points in many of their epistemological claims. As a consequence, the field is able to make visible and *extend* on an image of the Third World as Underdeveloped: no longer is it just lacking in agricultural technification, health planning and the like, it is also lacking in organizational expertise, managerial skills, human resource training capacity, efficiency, motivation and so on. It is a veritable list of all that is now additionally problematic, and a comprehensive guideline on what to treat and address as a result of this additional visibility.

The Approach to Knowledge

When looked at from this situated perspective, international and cross-cultural management appears as a cultured knowing, and a very particular "culture of knowing" that are functional and objective in its approach to knowledge and constitution of reality. Nowhere is this functionality more evident than in the field's treatment of culture. Contrary to the larger epistemological and historicized conjuncture highlighted by deconstructionists and postcolonial writers, culture in this discourse is reduced to the status of *a variable* (e.g., Burrell and Morgan, 1979; Morgan, 1986).

Furthermore, culture tends to be treated as a *secondary* variable, *exogenous* to the primary phenomenon under scrutiny (e.g., Kane and Ricks, 1989, 1988; Kanungo and Wright, 1983; Kotabe and Omura, 1989; Putti, Aryee and Tan, 1989; Tayeb, 1987). For example, consider the following claim by Stephens and Greer in explaining the boundaries of their research enterprise and interest:

> [W]e conducted a study of managers and professionals involved in [a] Mexican-US joint venture. Our goal was . . . to gain firsthand knowledge and understanding of the ways in which cultural differences challenge or disrupt managerial interactions in joint ventures [and] to identify ways in which individuals and organizations can circumvent or overcome these challenges (1995: 39).

The primary interest is not so much culture, but how culture *as* an independent variable can potentially affect joint venture management or success. The general intent here is not so much culture as it is to assess how culture as a general artifact can impact on whatever issue the researcher is interested in.

Moreover, not only is culture secondary, what this discourse of cross-cultural management addresses as "culture" is already by definition, *knowable*. It is a knowing that, as with the constitutions of subjects, is made possible and mediated by concepts already established in the field. Accordingly, constitutions and features of culture which authors identify are equally filtered by standard notions like motivation, commitment and uncertainty avoidance; and by interests in decision-making, efficiency, management capacity and so on.

As a variable, culture is often constituted as a potential "obstacle" and/or "integrative mechanism" that can aid in the effective transmission of modern management know how, and efficient administration of organizations in the international arena (e.g., Adler and Jelinek, 1986; Harpaz, 1990; Hofstede, 1987, 1985; Kedia and Bhagat, 1988; Lane, DiStefano, and Maznevski, 1997; Negandhi, 1985; Putti, Aryee, and Tan, 1989; Ronen and Shenkar, 1985; Tayeb, 1987; Usinier, 1991). According to Lane, DiStefano and Maznevski (1997: 2): "[c]ultural differences, if not understood, can be significant barriers to the implementation of a business venture." Or as noted by Stephens and Greer (1995) in establishing why "understanding cultural differences"(: 39) are important:

> [T]he management literature is unequivocal about the difficulty of establishing joint ventures. . . . [c]ross-national alliances almost certainly lead to conflicts when deeply held cultural assumptions initiate or compound differences in organizational processes, technology, and other

> factors. *Cultural values often lie at the heart of* these challenges, making it difficult to resolve the problems. . . . *[I]gnoring* culturally based challenges can be severe; they may include extended periods of poor performance, disharmony, missed opportunities . . . even the dissolution of joint ventures. . . . Thus, insights into cultural differences *can help* managers develop unique and effective ways to meet the challenges. . . . Cultural values can affect decision-making, managerial style, interpersonal trust, teamwork . . . (: 39–40, emphasis mine).

To these authors, understanding cultural difference is an important component of managers' ability to manage well and effectively in the international terrain.

Even in instances where researchers may start out by noting the nebulousness of the concept of culture and/or nationality (e.g., Adler and Graham, 1989; Boyacigiller and Adler, 1991; Hofstede, 1992, 1987; Kedia and Bhagat, 1988), whole countries and societies are still subsequently described in terms whose significance lie in their ability or capacity to say or show how such terms are *linked to* concerns with performance, productivity and efficiency (e.g., Hofstede, 1987, 1985; Ronen and Shenkar, 1985; Schneider, 1989; Usinier, 1991). According to Putti, Aryee and Tan (1989), interest in culture is important not only for its influence on commitment and work values, but because commitment and work values are related to certain behavioral outcomes such as absenteeism, retention and performance associated with the economic health of organizations. The economization of reality noted previously anchors now constitutions of the social.

From the standpoint of institutional ethnography, these conceptually mediated identifications and specifications regarding culture are part and parcel of ways through which fields of knowledge are able to constitute for themselves the specific terrain of their knowledge claims and visibilities. Going back to Stephens and Greer (1995), the cultural features they identify such as people's "assessments of those with whom they work" (: 40), their "likel[ihood] to challenge or oppose a supervisor's ideas" (: 41), "hesitat[ion] to provide decision-making input" (: 41), and "adapta[bility] to teamwork" (: 48) are all part of an effort to identify how—and why—culture matters. To the degree that these authors are able to do so, they are actively involved in *establishing the parameters of culture* as a variable. That is, what it is composed of, and how or where to see it.

A feature axiomatic of reductionist science (e.g., Burrell and Morgan, 1979; Morgan, 1986), these identifications are a procedure by which variables of interest to a field are established and elaborated on. Once established, they provide a plethora of concepts, causalities and *lenses* by which *other researchers elsewhere* can further explore and extend the claims of this knowledge. Referring back to Stephens and Greer (1995), their description of culture and their specification about its features not only resulted in the active linking of culture to joint venture management, it also resulted in further elaborations connecting culture and "managerial style" (: 40), culture and "decision-making processes" (: 41), culture and "teamwork" (: 48), and so on. It is to this elaborative procedure that Escobar (1994b, 1988, 1984) refers when he argues in his deconstruction of the development sciences that modern fields of knowledge have been effective social forces in their ability to *progressively* rationalize and organize the realities of developing nations, often setting into play new forms of knowledge, and techniques of power over which these nations must subsequently negotiate.

In sum, this process of (1) establishing that culture "matters," (2) elaborating on its parameters or features, and (3) identifying its links to other variables, is a central procedure in how fields of knowledge establish themselves *as a* field via the formulation or elaboration of variables. Framed and organized by the concepts, interests and forms of knowing *already* existing in the "theoretical armamentarium of the discourse" (Smith, 1987: 122)—that is, its institutionalized character—this process and practice of knowing anthropologize all subsequent encounters between researcher and reality, self and other.

Key to this process and possibility is the previously mentioned existence of "schemata" and "structuring procedures." To institutional ethnography, this creation and classification of categories undertaken by Stephens and Greer (1995) or other writers like Hofstede (1987, 1985) are very much involved in the "production of facts" (Smith, 1987: 73): in this case, the "fact" of culture. To Smith, such categories themselves say less about the "actualities of a naturally existing world" than they do about the "text-based methodologies" (1987: 153) and practices of academe organizing this world and our subsequent apprehensions of it. Of particular note to this "naming procedure" is how researchers' assumptions about, and reliance on an objectivist epistemology (Burrell and Morgan, 1979; Morgan, 1986) is precisely what is allowing for this possibility of reducing culture to a variable.

As an example, we find in Geert Hofstede's (1987, 1985) writings, the systematic identification and reduction of cultures the world over into four distinct categories: individualism-collectivism, power distance, uncertainty avoidance and masculinity-femininity. As he notes in his explanation on how and why such a categorization is possible even in the face of differences within nations:

> . . . [while] differences in culture within nations . . . do exist, for most nations we can still distinguish some ways of thinking that *most inhabitants share* and that we can consider part of their national culture or national character. National characters are *more distinguishable* to foreigners than to the nationals themselves (Hofstede, 1987: 311, emphasis mine).

To him, while differences exist, similarities [*what inhabitants share*] are still possible to identify. This is important since similarities presumably, outweigh differences as features by which Others and nations can be codified. Moreover, this ability of researchers to identify and codify is possible since the Other's character is "more distinguishable" to the foreign researcher and observer *than it is to themselves.*

Such active categorizing or "clustering" (Ronen and Shenkar, 1985) effectively reduces the heterogeneity of cultures into homogenized wholes. As an example, consider the following description of cultures by Hofstede (1987) using the conceptual lenses of his cultural dimensions:

> In countries with higher Power Distances—such as, many Third World countries, but also France and Belgium—individual subordinates as a rule do not want to participate. It is part of their expectations that leaders lead autocratically, and such subordinates will in fact, by their own behavior make it difficult for leaders to lead in any other way. . . . If the society is at the same time Collectivist, however, there will be ways by which subordinates in the group can still influence the leader. This applies to all Asian countries (: 321).

Not only are whole nations like France and Belgium, and regions like the Third World and Asia, lumped together according to whether they are high or low in power distance and collectivism, there is little by way of an acknowledgement over the specificities locating and distinguishing these nations and regions. Instead, generalized explanations and projections about their proclivities and tendencies are categorized as a given.

This creation and reliance on categories, like conceptual filters, silence the living complexity of Others whom we study. Not only do existing concepts like motivation, productivity and so on become *generic* or universal categories, cultural dimensions noted here also take on the same guise. They become ways of representing and reducing what may be diverse and shifting configurations within societies to an essential whole, often at the level of the nation state. It is precisely such procedures that are producing the cultural and economic reductions criticized by deconstructionists and postcolonial writers.

Of particular importance to this reductionist possibility and universalism is the practice of rendering the objects of study *external* to the observer (e.g., Burrell and Morgan, 1979; Escobar, 1994b, 1988, 1984; Prakash, 1990; Said, 1989, 1979). Encoded in Hofstede's (1987) ability to identify and categorize is a presumption that enough of an objective reality exists even in the terrain of the social to allow for this act of objectification. It is a presumption reinforced by an explicit privileging of the researcher's ability and capacity *to know.* While Hofstede's claim to know may, under certain circumstances, be a commentary about the author's non-reflexivity about his own cultural lenses, here it is an index of the politics of representation, and the relations of knowledge and power that allow for such a lack of reflexivity to go unexamined. As noted by Burrell and Morgan (1979) in their assessment of this objectification procedure:

> Theorists . . . tend to assume the standpoint of the observer and attempt to relate what *they* observe to what *they* regard as important elements in a wider social context. . . . The conception of science which underlies the paradigm emphasizes the possibility of objective enquiry capable of providing true explanatory and predictive knowledge of an external reality . . . [while also] . . . attribut[ing] independence to the observer—an ability to observe what is, without affecting it (: 107, emphasis mine).

Practices of objectification come already embedded in, and socially organized by the scientific claims informing how and why researchers know. Specifically, science and scientific authority itself, along with assumptions about the existence of an objective reality, are what is organizing the claims of these constructors to know, and their location as *knowers* in relation to their respective objects and subjects of inquiry. This assumption of an all-knowing observer set apart from the object and subject of study forward to readers a constitution of Others as objective and universal *fact.*

When questions regarding difficulties attached to knowing in a cross-cultural or international terrain are asked, such questions are often reduced to the problem of *method.* According to most researchers considering such difficulties, it is primarily a question of "accuracy," "measurement," "validity," "comparability" and other concerns of a primarily methodological nature (e.g., Adler, 1984; Adler and Graham, 1989; Adler and Jelinek, 1986; Adler, Campbell and Laurent, 1989; Boyacigiller and Adler, 1991; Hofstede, 1987; Jorgensen, 1979; Kanungo, 1983; Nasif, Al-Daeaj, Ebrahimi and Thibodeaux, 1991; Rieger and Wong-Rieger, 1988; Roberts and Boyacigiller, 1984). As noted by Roberts and Boyacigiller (1984):

> [P]roblems in . . . investigations [include] . . . weak conceptualization and poor linkage from conception to operation . . . [the] way measures are obtained, aggregated, and analyzed . . . [l]evels of analysis. . . . [and]

> problems concerning time . . . including lack of justification for time intervals over which data . . . are collected . . . or are collected using very different "envelopes" of time . . . studies looked [at] . . . are almost non-integratable [providing a] serious problem in terms of pushing forward future research (: 460–461).

Thus, difficulties attached to researching cross-culturally lie primarily in a plethora of methodological challenges ranging from how researchers operationalize their concepts to specific problems connected with data collection and analysis.

Nowhere within this constitution of the problem is there any questioning over the nature and feature of the assumptions or know how informing the production or accumulation of cross-cultural knowledge. Rather, concerns are organized around an internal discussion over how to better measure, or validate the concepts and methods or methodological procedures up-holding the research and its claims. What *are* unquestioningly up-held are the assumption of an objective reality, and the capacity of researchers to know. The presumption here: the Truth of culture—and Others' culture—is there for us to identify and to know if only enough consistency and methodological vigor exists.

Thus, researchers' concerns over method are directly attached to the claims of the field *as a science;* their preoccupation with method attached to a need to *claim* scientific authority. As noted by Burrell and Morgan (1979), conventional practices of scientific inquiry shift the authority of writers and researchers onto the shoulders of the methods upon which the research enterprise itself rests. So that in effect, the efficacy of both research and researcher as knowledge and expert rest on the legitimacy and authority of the tools, methods and schemata they rely on and use.

It is this silence between "culture as a concept" and the "cultured-ness of concepts" that simultaneously produces and reinforces the universality of cross-cultural management knowledge, and the assumption of an identifiable Other. In the process, the cultured lenses constituting the researcher-observer's visibilities are hidden: the values and assumptions informing their sight rendered invisible under the guise of descriptions or categories, and upheld by the conventions and practices of science and scientific authority—even as these assumptions and values are *precisely* what make possible the conveyance of some shared understanding between researcher and field, writer and reader. In effect, what is produced is the presence of a knowable Other, and a silence about the cultured Self.

In this regard, economic reductions not only help produce and confirm the underdevelopment of those we study, they are what produce this silence behind which the cultured Self hides. Indeed, as an act or a process that provides the uniformity and stability needed for *comparative* research to be conducted and knowledge generated, these reductions make possible the very existence of international and/or cross-cultural management as a field of study, through in this instance, the provision of a possibility and *ability to* compare, cluster and grade nations (e.g., Hofstede, 1987, 1985; Ronen and Shenkar, 1985). This economism is thus a key condition of discursive possibility. Put another way, it is a practice and a procedure by which the Underdevelopment of the Third World is rendered *visible, measurable, comparable* and most of all, *real.*

SUMMARY

The foregoing review and analysis show the culturally and socially situated feature of our knowledge about and on Others: namely, modernist, reductive, organization-centered and linked to capital. By focusing on the social organization of how this knowledge is constructed, it turns our attention to the practices and relations of knowledge and power informing this possibility. From the standpoint of institutional ethnography, this attendance to the cultural heritage and practices informing the claims and interests of the field is needed to problematize and re-situate what we take for granted about the objects of our knowledge. Within this purview, the possibility of "identify[ing] *culture-free* theories of management" (Doktor, Tung and Von Glinow, 1991a: 259–260) is oxymoronic since it is precisely the culturally specific nature of our theories that ensures against the possibility of having culture-free theories. The ways we construct knowledge under the norms of universalism *already precede* all our knowledge, and thus renders it culture-specific to the West.

Moreover, by attending to these intersections between culture, knowledge and practice, the review brings into sight *the presence* of a researcher and knowledge constructor who is centrally engaged in those activities and practices which produce and socially organize what we accept as knowledge. Seen from this perspective, these individuals and other like knowers are crucial *linking pins* and *social sites* mediating between values and knowledge, thought and action, profession and convention, knowledge and practice, and self and other [see Figure 3.1]. As owners and carriers of culture, knowledge and practice, they embody, in and through their [social, cultural, physical, epistemic, professional] presence, the histories, contexts and training they have been subjected to. Accordingly, when they research and write, these individuals do so mediated by the specific locations situating them as knowers and doers.

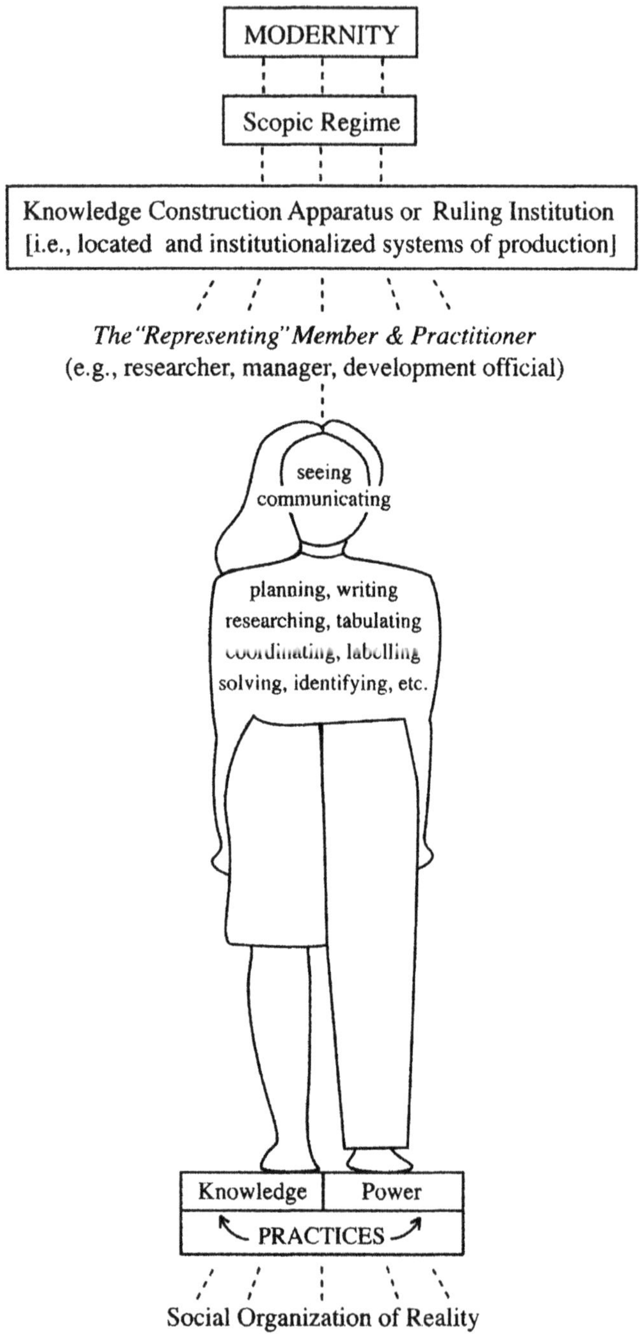

Figure 3.1 Knowers as Mediators of Culture, Knowledge and Practice

Put another way, researchers are not just researchers; they are active *agents* and *subjects* of knowledge and power. But as noted in chapter two, it is not just researchers or knowledge constructors who are agents of knowledge and power. As I seek to show in upcoming chapters, the managers, officials and workers I talk to and observe during my time in the field are all producers and users, agents and subjects of knowledge. By *re-situating* and *re-locating* what people do and how they know in terms of the systems of knowledge and power they inhabit or are located in, institutional ethnography provides a way of accounting for how everyday practices of individuals in their respective institutions are constantly involved in socially organizing reality via the textual and documentary forms, professional norms, and so on noted previously. This endpoint to the discussion above is my starting point of analysis; the *theoretical entry-point* into the *empirical terrain* of Malaysia as development case study and living experience.

Chapter Four

Towards Vision 2020 and the New Malay[sian]

At the point of my entry into the nation for fieldwork on this project early in 1995, Malaysia, as nation and economy, was already well on its way to becoming part of what multilateral agencies like the World Bank (1993a) calls the "East Asian Miracle." This term generally refers to that compendium of newly industrializing nations flanking the eastern edge of the Asian continent and the western shores of the Pacific Ocean (Singapore, Korea, Hong Kong, Malaysia). Hailed for their receptiveness to economic liberalization policies, stable macroeconomic and political environments, and general effectiveness in transforming their respective local economies into forces of global participation and action (e.g., World Bank, 1993), these nations were touted and held up as prototypes to developing nations elsewhere like Latin America (e.g., Edwards, 1995; World Bank, 1993a). According to these reports:

> [t]he market-friendly approach [best] captures important aspects of East Asia's success. These economies are stable macroeconomically, have high shares of international trade in GDP, invest heavily in people, and have strong competition among firms (World Bank, 1993a: 10).

Heralded thus, these nations were instances of the viability of a particular development policy long advocated by these agencies: namely, the positive-ness of open market policies in general, and the efficacy of structural adjustment policies (SAPs) in particular. To advocates of this market-friendly approach, "rapid growth" is mostly associated with "carefully delimited government activism" (World Bank, 1993a: 9; see also *World Development Report 1991*). Where "governments generally have failed," they fail because of their reliance on means of "resource allocation . . . *other than* market mechanisms" (: 10, emphasis mine).

Reduced to its most basic, the above can be read as the following: development is about the ability of nations to adopt neoliberal policies. In the process, nations like Malaysia are codified in terms of their ability or capacity to make such adjustments and little else. As noted in the two preceding chapters, absent within such tales of transformation is a sense of the sociological universe surrounding the constitution or deployment of these policies. Similarly, little attention is paid to local exigencies beyond the economized calibration of nations: that is, their competitive climate, investments in people, administrative efficiencies, labor productivity or skills and the like.

Framed within this official story of development, the aim behind this chapter is twofold. On the one hand, it is intended to assess how neoliberal discourses of modernization are changing both the development terrain and the constitution of [under]-development in Malaysia. Policies formulated and adopted by the Malaysian government since the mid-eighties form the backdrop of the current analysis. Particular attention is focused on how this "market-friendly approach" is contributing, through its specific representations of local problems and issues, to the production of a very particular form of knowing which sees the transformation of local subjects as crucial for effecting the transition of nation and self to developed entity. This focus on the mediation of local visibilities and actions by modern market discourse renders understandable the very policy decisions adopted by local state planners, not to mention the impetus for and interest in transfers of knowledge forming the entry-point of this project.

On the other hand, however, this description of development as applied to the context of a Malaysia in the midst of change is only a part of the story on hand. The adoption or constitution of such policies is neither sealed-off from the exigencies of the local culture and terrain, nor is it absent of history and accountability, whether local or global. In Malaysia's instance, the specific forms of visibilities encoded within development discourses are actively appropriated by the state in response to very particular issues and concerns specific to its locality. In the process, these representations become effective social forces enabling and providing local state planners with the ability to actively reconstitute problems and subjects that allow for their subsequent action and intervention in the local arena.

Implicit within the above is a commentary about the simultaneous process of constitution and appropriation *that is* the transfer process as it is played out in the local terrain. It is a process mediated by culture, by history, by memory and by an Other who is neither passive recipient nor *tabula rasa*. Though admittedly, the mediation of change *at the local level* by certain institutional practices and everyday work procedures of ruling individuals and discourses of development and management are, as I seek to show in subsequent chapters, power encoded activities whose immediate effect is the transformation of sight and so, of action. Although even here, processes of constitution are neither accepted unquestioned nor unmediated by local concerns. Thus, contrary to the unifying and linear assumptions inherent within mainstream understandings of how western and modern forms of know how are transferred (e.g., Edwards, 1995; World Bank, 1993a, 1993b), objects of development like Malaysia are *simultaneously* the subject[s] and active agents of such knowledge, often contributing to the reconstitution of dominant discourses even as they are simultaneously constituted by them.

THE TRANSITION TO NEOLIBERAL POLICY AND MALAY BACKWARDNESS

Partially triggered by the worldwide recession in the mid-1980s, the Malaysian government began a concerted attempt in 1984 to structurally adjust the economy away from a focus on government to private sector investment as its "engine of growth" (Mohamad, 1993a: 424). The collapse of world commodity prices for its mainstays (e.g., tin, rubber, palm oil), technologically-driven discoveries of synthetic and natural substitutes for certain commodities in the late seventies and early eighties (e.g., Jesudasen, 1989; Mohamad, 1986), and the "unsustainable" and "dismal" (Mohamad, 1993a: 423) results of certain "non-financial public enterprises" had plunged the economy and nation into large scale debt. Rather than a commentary about the struggles newly de-colonizing nations like Malaysia must go through given a political economy whose dominancy is linked to close interpenetrations between capital and science (e.g., Escobar, 1994a), these conditions were taken as *evidence of the inefficacy* of regulated development policies and hence, the superiority of market driven models of development (e.g., Edwards, 1995).

Compelled by falling exports, collapsing commodity prices, declining government expenditure, and the repayment of debt to restructure, the Malaysian government actively made a bid to transition towards neoliberal development policies. Relying on a mix of policy instruments and the centralized capacity of the state to oversee, implement and ensure that regulatory policies governing financial, economic and administrative arenas were set into place, what followed was (1) the progressive deregulation and privatization of certain government agencies and functions, and (2) the increasing reliance on foreign direct investment (FDI) and export industrialization as generators of growth (Mohamad, 1993a, 1993b; World Bank, 1993a).

Seen from the perspective of this market driven approach to development, the nation's current development blueprint, the New Development Policy (NDP; 1991 on)—popularly known as "Vision 2020"—is an extension and elaboration of this basic strategy of economic conversion [see Appendix B for a diagrammatic summary of these development objectives and strategies and their links to Vision 2020]. Through the active fostering of human resource development activities, promotion of science and technology, continued deregulation of government agencies, and creation of an active and viable Bumiputera Commercial and Industrial Community, the intention of the government is to create a base from wherein the nation and populace is "equipped" for a development founded on administrative and technological proficiency, innovation and export industrialization (Mohamad, 1993b; Mohd Sherrif Mohd Kassim, 1993). Growth, rather than coming from the state will be led primarily by the private sector. Also included is the continued, albeit reduced reliance on the national affirmative action policy initiated under the New Economic Policy (NEP; 1970–1990) to elevate Bumiputera[1] participation and representation in all modern sectors of the economy.

However, as noted earlier, this description of development prescriptions as applied to the context of a Malaysia in transition is only a part of the story. Central to the formulation of the nation's development policies since its independence is a concerted effort on the part of the state to both deal and come to terms with a national polity who is neither homogeneous nor even in its social, political and economic make up (e.g., Crouch, 1996; Hua, 1983; Husin Ali, 1984; Jesudasen, 1988; Jomo, 1986; Kahn and Loh, 1992; Khoo, 1995; Ratnam, 1965); a state and condition of detente due in no small part to the nation's colonial history. In particular, the specific connection between underdevelopment and the Malay community, and how such a connection may be "resolved" has been and remains a central concern galvanizing the actions and decisions of the state (Mohamad, 1982). This tension engendered by the pluralistic heritage of the nation was very much a factor in the time of my childhood in Malaysia and it still was very much a factor during my return to the nation for this project, albeit one articulated in somewhat different ways under somewhat different conditions.

From the discourse of poverty and the essentialist focus on race forwarded by Mahathir Mohamad (1982) during the NEP years to the NDP's focus on self sufficiency, economic resiliency and industrial discipline, what we see as we trace through the changes in the state's discourse on national development is an active attempt to *reconstitute the divisiveness* engendered by the NEP's singular focus on poverty and race.

Through the active appropriation of market discourses by state officials, focus is trained instead towards the construction of a progressively unified, equal and ever-productive and "resourceful" national polity. It is a discursive and policy re-orientation whose intention is to override race and poverty *in favor of* the connected possibilities attached to the transition of nation and citizenry to a developed and modern entity. A re-orientation where the rule of the market will confer on all an implicit equality and level playing field, at least in so far as the economic terrain is concerned. As I was to realize as my fieldwork progressed and my apprehension of the complexities attached to Malaysian society and history became an equal part of my research time, to even begin understanding this transition from a singularly Malay focus to a more generally "inclusive" Malaysian emphasis entails a *prior* understanding of the meanings and struggles indexed by the emergence and the formulation of the NEP itself; of how it came to be as an official policy of the state.

Central to my analysis here are the writings and positions of Malaysia's recently retired Prime Minister, Dato Seri Dr. Mahathir Mohamad (1981–2003). As noted by Khoo Boo Teik (1995; see also Khoo Kay Jin, 1992) in his intellectual biography of Mahathir Mohamad, he has been and remains for most Malaysians, whether Malay or non-Malay, a central figure in the life and times of post-independence Malaysia, both during the communal-driven strife leading to the formulation of the NEP, and the post-NEP years, particularly so in his role as "visionary" of a developed Malaysia. Both articulator of national change and critic of western economic and political actions, we see in his evolving constructions of Malays and nation alike, the interpenetration of locally driven concerns and globally embedded notions of [under]-development—along with the tensions and paradoxes engendered by this sometime uneasy and synchronic relation.

* * *

I think about the many journeys that Malaysia and all Malaysians—including myself—have gone through. Memory like the broken mirror that Salman Rushdie (1992) talks about comes automatically to mind, and I think, yes, I suppose we are all broken mirrors in one way or another when it comes to memory. Yet memory is invaluable, even possibly more precious, as Rushdie contends, precisely because of this: a nation and its sacrifices; the search for identity; constructions and definitions of Malayness and nation-hood; the how-to of solidarity and common grounds; tea and coffee in little kampung houses; the silhouette of a father keeping guard; even the checking of water-flow in the dark of night. All these images vie with one another for preeminence in my mind as I think about my own journey and how it has contributed to the constitution of the present "I." And a sense of loss and nostalgia besets my consciousness: the could-be's and the might-have-been's, and what is and was.

Lessons in nation-hood? And colonialism? How much more precious and significant such events are when filtered through the lens of the personal observer. For then meaning comes so clearly identifiable. Lives and choices shaped and made on an everyday basis—so much more alive than 1,000 page-tomes on official policy and statistics. Yet, as Rushdie and Escobar and Chatterjee show in their writings, all the more poignant because of the capacity of this distanced knowledge and these objective representations to affect how lives are to be lived and choices made. On the other hand, non-science [whatever and however this may be constituted] seems to consistently have the last laugh as it sits back and observes how our rationalist world of bureaucrats, scientists, economists, politicians, anthropologists and planners scramble to try and figure out why and where they could have gone wrong as unintended consequences abound, and actions aimed at some desired end leads to further undesired beginnings.

So much for domination over structures and processes. This "virtual reality" of control and functional equilibrium takes on even greater urgency as we go on to consider the magnitude of how policies and decisions by experts and powers-to-be have changed people's lives and constitution of selves—here, there and everywhere. Not so funny after all. To quote Juliet Schorr (1992), how "to exit the squirrel cage"?

* * *

ROAD TO THE NEP: PLURALISM AND THE MAY 13 RIOTS

Malaysia is often held-up by political and social historians as "a classic case of the plural society" (Crouch, 1996: 13). Composed of a multitude of ethnic communities, including Malays, Chinese and Indians, these delineations are paralleled by almost equally clear lines in terms of the communities' social and economic differences: a veritable amalgamation and intersection of locations marking each and every community. As noted by observers of Malaysian society (Crouch, 1996; Ratnam, 1965), socially, all Malays are Muslims, whereas the Chinese and Indian communities are more fragmented along religious, linguistic and cultural lines. Economically, Malays were predominant in the bureaucracy and agriculture, while non-Malays predominated in commerce, industry, the professions and the working class.

At the time of its independence from the British in 1957, the population of Malaya was almost evenly divided between the Bumiputeras and the immigrant communities (Crouch, 1996; Ratnam, 1965). Although the entry of Singapore and the states of Sabah and Sarawak leading to the formation of Malaysia in 1963 changed this configuration somewhat, the subsequent expulsion of Singapore from the nation in 1965 "tilted" the balance yet again in favor of the Bumiputera community (Crouch, 1996). By 1970, Bumiputeras made up 55.5% of the population, Chinese 44.5%, Indians 9.0%, and others 1.4% (Government of Malaysia, 1981: 74).

As a result of this almost equal division in demographic make-up between the Malay and non-Malay communities, along with the seemingly clear-cut differences in locations that index each community, virtually all social issues and policies—political, economic, cultural—in the nation were and are constituted in communal terms. As Crouch (1996) describes it, almost all policies "were seen as benefiting one or another of the main communities, while anything which benefited one community tended to be seen as depriving the others" (: 14).

It is within this pluralistic and communally tinted backdrop that events leading to the formulation of the NEP must be understood. The result of an accumulation of communally-charged events and tensions throughout the 1960s which culminated in the communal violence of May 13th, 1969 (Crouch, 1996; Khoo, 1995), the riots were triggered by election results signaling a profound crisis of discontent amongst both Chinese and Malay voters alike. In particular, the unexpectedly strong showing by primarily community-based opposition parties countered a commonly-held belief in the post-independent and "alliance-bound"[2] nation that the "stable management of Malaysia's plural society" can be based on the "'separatist formula" of "Malays in politics" and "Chinese in economics" (Khoo, 1995: 26). Over the course of two and a half months or so, 196 people were killed, approximately 75% of whom were Chinese, although unofficial estimates suggest that the number of Chinese killed is much higher (Crouch, 1996: 24).

* * *

Being all of five years old or thereabouts, my apprehension of May 13, 1969 is confined to my immediate surroundings and everyday activity and routine. Absent, I suppose, as some might point out, of knowledge about and apprehension of the history, politics and racial undercurrents informing those events and a time when communal riots broke out all over the country. Setting into play much of what would eventually be defined as "Malaysia" and "Malaysian."

We lived by the edge of a water village in Kampung Sembulan, Kota Kinabalu in a big wooden house with concrete pillars and verandahs at both ends of the house. The space beneath the house, stoutly held up by the aforementioned pillars, consisted of a huge tract of partially reclaimed land that extended into the sea. On this land were our various trees, garage, storage space, fishpond [somewhat like a big ground-level pool], "make-shift" basketball court and orchid greenhouse. Also to be found on occasion, a (speed)-boat or two, and my mother's coterie of chickens, ducks and whatnot. The area behind our house consists of rows upon rows of neatly-lined houses, stilts rising from the sea, connected to each other by an orderly series of wooden bridges which constitute our neighborhood's roads and alleyways. Many of these were fairly grand affairs like ours, built primarily from wood, nicely painted, some with concrete steps and shingled roofs.

When the rains come, our "field" would be flooded. Once when I was quite young, I remember waking up halfway through a particularly stormy evening during the monsoon season to find my brothers, cousins and their friends paddling crazily and joyously in the "shallows" which used to be our basketball court while the grown-ups were just as noisily and enjoyably playing mahjong inside.

* * *

For the Chinese, increasing economic disenfranchisement within the community's urban working class, the legacy and influence of Singapore's People's Action Party calling for a more inclusive "Malaysian Malaysia," and increasing doubts about the ability of Chinese leaders within the Alliance to represent the community's interests led to the massive turning away of support. Malays, on the other hand, were increasingly concerned about the status of their sovereignty as "true sons of the soil." Singapore's entry into the federation in 1963 had, however temporarily, increased the proportion of Chinese in the nation. This, along with calls for greater inclusiveness provided what seemed to be indelible challenges to their political status as guaranteed by the independence principles of 1957. Concurrently, the decision of the state to keep English and "maintain indefinitely" the implementation of Malay as the national language led to a growing conviction within the community that Malay leaders were too moderate and lenient, more intent on maintaining the status quo or upholding their aristocratic[3] interests than in advancing the causes of the community (Crouch, 1996; Mohamad, 1982). More than anything however was the increasing frustration of Malays regarding the lack of their economic progress.[4] Thus, as in the Chinese community, there was a crisis of credibility within the Malay community over the efficacy of the Alliance as a social, economic and political form.

* * *

To the northern end of this predominantly Chinese enclave was the Malay portion of the kampung. The houses there were somewhat smaller, sometimes colorfully painted, sometimes not. All were connected to the more Chinese portion by the aforementioned network of bridges. I remember going for visits there on numerous occasions, tagging along with my father on his visits with friends. Pakcik Awang in particular was a favorite of the family. Somewhere between these two neighborhoods lived some friends of the family whose father was an ice-cream maker. To this day, I still have extremely fond memories of his cream of corn ice cream. There was a phase where I would wait religiously every day at a certain time of the afternoon to hear the tinkle of his bicycle's bell. As he passed by on his way home, I would stop him and make my twenty cents' worth of purchase. He often gave me just that "extra-bit" which always made my day.

Straddling between the land and sea, our house had a bumpy dirt and pebble lane leading to the main road into and out of the village. Situated between our house and the single row of then very modern concrete shop-houses and market were a number of like "straddlers." Some relatives lived atop one of these shop-houses, as did numerous others in stilt-houses back of the shops and market. These shops, they were a veritable storehouse of childhood bric-a-brac. One in particular was my favorite. Anything and everything it seems could be bought there: from watercolor sets to pens to candy to ice cream to paper dolls and drawing paper. Everything. Even Cadbury chocolates. Next to this magical storehouse was an open-air coffee shop whose beef noodle soup, ice-kacang and homemade toast with gaya and butter was the "absolute."

Many were the lazy afternoons when I would venture out [when I was deemed old enough that is] and come home with my precious toast and ice-kacang in tow. The journey between our house and the stores was unfortunately quite treacherous at times, peppered as it was by quite a large number of strays and flocks of geese belonging to one of the neighbors. Once I made the mistake, upon noticing the "gleam" of some geese's eyes, of running for home as fast as my legs could carry me [which was not very fast]. Needless to say, the evil creatures came after me with a vengeance and speed that impressed me a great deal as I sped home. To this day, my motto remains: never look geese in the eye and run only when you think you can very safely and easily out-distance them. It is an eminently wise motto should anyone wish to put it to a test. My childhood and knees were constantly marked by bruises and plaster [and holes in the case of my leggings] for a good reason: a bad eye and an inability to walk or run properly. Either that or I was just accident prone.

* * *

From the perspective of a postcolonial reading of this official story, what is contested above are issues of representation and voice within both the Chinese and Malay communities, albeit one contested on different levels and for different reasons depending on location. Differences in history and experience, evinced here by different concerns voiced by members of the respective communities—for example, inclusiveness and voice for the Chinese; potential silence, aristocratic interests of leaders, and economic marginalization for the Malays—highlight the difficulties of representing what is an ostensibly heterogeneous polity via the reliance on an elite (Malay) or moderate (Chinese) few. Differences that call into question the assumed homogeneity [in voice, in experience, in history, culture, language] that exists within the clearly demarcatable boundary of the nation state or the community, and the impossibility of representing such diversity either politically or otherwise.

* * *

> *This meandering journey through a bygone era with long-gone neighbors is necessary in some ways to juxtapose what life was like for me during those days before it was so glaringly interrupted by events older and bigger, it seems, than five-year-old me. An interruption which is admittedly temporary in terms of the temporal spectrum and reality that is a child's world, although the memory indexing and forces underwriting the event certainly extend beyond the child's immediacy of comprehension.*

* * *

The NEP and the Malay Dilemma: Constructing the Basis of Nationhood and Self

It was in this atmosphere of national uncertainty and communal negotiation that Mahathir Mohamad, labeled an "ultra" (radical Malay nationalist, Khoo, 1995) because of his vocal critique of the government and its detached concern for the community, wrote what would eventually be identified as the "definitive statement" of Malay sentiment and frustration during this time:

> . . . not only is there little effort made to right the economic wrongs from which they [i.e., Malays] suffer, . . . [t]he whole idea seems to be that the less they [i.e., Malay leaders] talk about it the more the country will benefit from the economic stability built on Chinese economic domination. What is important, the Malays are told, is that Malaysia must prosper as a nation. . . . The Malay dilemma is whether they should stop trying to help themselves in order that they should be proud to be the poor citizens of a prosperous country or whether they should try to get at some of the riches that this country boasts of, even if it blurs the economic picture of Malaysia a little (Mohamad, 1982: 60–61).

Written to protest the ways by which Malay concerns are systematically silenced in the interest of "economic stability" and "national prosperity," *The Malay Dilemma* (1982, first published in Singapore in 1970) is important not only for understanding what has been called a crucial turning point in the nation's history, it is also significant for what it says about the often paradoxical and ironic feature of how nations like Malaysia must rely on a language not their own in order to give voice to a self.

Constructed under the tropes of poverty and race, the primary intention of the book was to highlight the "inferior economic position" (Mohamad, 1982: 32) of the Malays and suggest reasons for why this might be so. Relying on very particular constructions of both Malays and non-Malays (i.e., Chinese) alike, the book argues for greater "constructive protection" (Mohamad, 1982: 31) by the state in order to effect the "rehabilitation" (: 103) of the Malays.

* * *

> *My first inclination of "difference" was a certain tension in the air as my parents, aunts, siblings and cousins talk about stocking up on food and whatnot. Many of our Chinese neighbors moved away temporarily from their houses and returned to their family homes in more rural areas where it was deemed safer, since these were neither lumped in the middle of a Malay village nor located in the center of towns. We decided to stay, and were joined by many of our cousins and friends. All came and stayed with us for a few days: packing rations, clothes, apprehension, laughter and kids.*
>
> *Registering the tensions around me, my first and most immediate emotion was one of joy since it [the exodus by all to our quickly expanding communal household] meant that I could have hours of fun with my young friends. What a treat. Yet we were never totally allowed to forget about the curfews and state of emergency, not to mention dissolution of relations with our longtime Malay neighbors and friends. This would often come in the form of direct admonitions not to go out of the house, not even so far as stepping out onto the balconied verandahs. Doors were always locked, windows sometimes closed. My father organized a rotating system of guards and "watch" for the men. Once, on waking in the middle of the night, I remember seeing him sitting by himself, smoking peacefully as he kept vigil over us. His silhouette lit occasionally by the tiny flame from the tip of his cigarette, a length of water pipe by his side. A silhouette, a flame and a water pipe: this was my May 13th, 1969.*

* * *

At the center of this need for protection and rehabilitation is the existence of a Malay subject who is singularly incapable and unprepared, for reasons of geographical abundance, rural isolation, heredity and culture, to confront either the increasing competition from immigrant communities or the vagaries of a modernizing society. In particular, the "abundance" of their environment, and the "inherent traits and character" (Mohamad, 1982: 31) of the community produced a situation where "the observation that only the fittest would survive *did not* apply" (: 21, emphasis mine).

Specifically, Malays' lack of "vigorous work" or "mental activity" (Mohamad, 1982: 22), their "neglect of children" (: 28), "in-breeding" and "abhor[ence] of celibacy" (: 29) were all held up as causes of "Malay economic backwardness" (: 57). Such was the situation of the Malays, whose location in rural areas and living conditions render them vulnerable to disease, mired in the ignorance and illiteracy induced by both physical and social isolation. Moreover, they were further "handicapped" by certain cultural characteristics inherent within the community. Namely, their so-called "docil[ity]," "letharg[y]," "self-effac[ement]" (Mohamad, 1982: 30), "unprotesting tolerance" (: 158), "politeness," "abhorrence of unpleasantness" (: 117), "natural courtesy" (:142), and "self-restraint" (:171). According to Mahathir Mohamad, the "Malay social code . . . can only lessen [their] competitive abilities . . . and hinder their progress" (: 171). Together, all served to produce a people who are not only "complacent [in their] reliance on others" (Mohamd, 1982: 30), but "ill-prepared to face the challenge of living in competition . . . with aggressive immigrant races" (: 30).

By contrast, "the Chinese who flooded Malaya" were "adventurous and resourceful" (Mohamd, 1982: 25); their "industrious[ness] and determin[ation]" (: 25) leading quickly to the displacement of Malays in petty trade and skilled work: "[w]hatever the Malays could do, the Chinese could do better and more cheaply" (: 25). Against the "easy-going . . . toleran[ce]" (Mohamad, 1982: 85) of the Malays is the "predatory," "hard-working and astute" (: 85) Chinese. A contrast in community features and [racialized] characteristics that almost always seemed to end in the economic domination of Malays by Chinese. At the center of this possibility for subjugation is a problematic Malay subject who is "disinclined to work" (Mohamad, 1982: 28), and "[u]nab[le] to understand" (: 167) the "value [of] time" (: 163), "the potential capacity of money" (: 167), and "real value of . . . property" (: 169). Or, as surmised by the future prime minister: "[i]n this sort of society, enterprise and independence are unknown" (Mohamad, 1982: 29).

What is being constructed here is a simultaneous situation of unequal competition and a constitution of Malays as individuals who must be actively "helped" *out* of their poverty, illiteracy, ease-of-attitude, complacency, non-economic dominance and so on. Both constructions rely on the constant juxtaposition between a backward and helpless Malay Self and a resourceful and aggressive Chinese Other.

> *I remember well the excited and somewhat frightened stories my cousins or siblings would tell whenever they come home from some quick jaunt or sojourn outside: the torch-carrying mobs, the exodus to Muslim graveyards, the ceremonies there, the clashes, the burning of Chinese warehouses and so on. Regardless of authenticity and breadth, these were images lurid for their codification by that mass of swirling emotions seemingly running rampant within the hearts and minds of its messengers—images that were subsequently built to ever-greater proportions by my youthful imagination. Suddenly, our former neighbors and friends from the Malay village did not seem so familiar after all, even as stories about "secret" communication between the Chinese and Malay villagers were also bandied around.*
>
> *I do not myself remember witnessing any houses being burned, or people fighting. But I do remember waking in the middle of yet another night and hearing my aunts and father and cousins whispering about "some men" who were checking to make sure that our tanks are full of water in the event of a fire. A peek at some point out of my bedroom window, which overlooks the fishpond, did show the galoshed silhouette of two firefighters checking to see that the pipes were running. In the dim light of the night, they looked doubly alien in their protective gear. Rather than comforting me, this only made me more nervous and uneasy. Mobs and whatnot I am not quite sure I apprehend, but fire, even the potential of one, certainly did register with me. My world did not seem quite as warm in its cocoon at that moment.*

* * *

More significant still to note is the degree by which such constructions of poverty and backwardness, and hierarchical differences between communities are *already* infiltrated by singularly modernist assumptions (regarding progress, evolutionary possibility) and values (competitiveness, timeliness) on the one hand, and produced through a reliance on equally modernist techniques of application on the other. It is an infiltration and a production that relegates to the self a position of inferiority and a definition as "problem to be solved" (i.e., Malay backwardness, Malay dilemma).

As noted by Escobar (1994b) in his deconstruction of development discourses and developing nations' encounters with development, implicit within the construction of underdevelopment which the concept and discourse of development establishes is not just the systematic unveiling of characteristics or features associated with or designated as underdeveloped, but also a [social] target such as a nation or a community towards whom actions geared at rectification are aimed. In other words, the active identification and construction of Malay backwardness forwarded by Mahathir Mohamad in his assessment of the Malay Dilemma is *already* part and parcel of a modern knowledge construction process whose effect is the production of client categories—like rural Malays—that are then subjected to treatment. As noted by Khoo Boo Teik (1995), Mahathir Mohamad's adherence to a Social Darwinist constitution of people and progress, an assumption axiomatic of evolutionary modernism, is central to his call for "constructive protection" [or is it "inoculation" given his training as a medical doctor?] *since* due to reasons of character and so forth, they are less equipped to succeed in a changing and ever competitive society without the aid of this institutional intervention.

* * *

A montage of images and a series of excited and frequently whispered exchanges—this was my May 13th, along with the joy of eating and playing with my gaggle of equally noisy friends and cousins. Yet the memory of my father's silhouette and men checking for running water in the night remains to this day. Untainted by time, sometimes with little feeling save for the unusual quality of the time, sometimes by a tinge of the fear that had shot through me at these silent activities in the night.

Memory like a broken mirror, a then and a now, a me and a you, an us and a them. Yet can they be dismissed as "beside the point"? Off the mark and unreflective of real issues and struggles going on at the time? In the state? Elsewhere in Malaysia? Even elsewhere in another place at another time? Somehow, I do not think so.

* * *

While primary attention has been focused here on Mahathir Mohamad's constructions of the Malay dilemma and the challenge it poses for long term national stability and unity, it needs noting at this juncture that the Malay elite were themselves increasingly aware in the aftermath of the May 13 riots that they were in danger of losing their base of support within the community unless they were more responsive to grassroots concerns. As noted by Crouch (1996: 24), the 1969 election "had shown a strong swing within the community away from the government," especially in those states the government "regarded as its own territory." The rioting had revealed pent-up frustrations not only amongst members of the community in general, but amongst young ultras like the future prime minister. As Crouch goes on to note:

> The Malay elite came to the conclusion that the moderate, gradual approach to racial questions exemplified by the policies of then Prime Minister Tunku Abdul Rahman was no longer sufficient to retain the large-scale support of the community. Instead, a new program had to be

> launched in which Malay political power would be used to bring Malays into the mainstream of the modern economy and thus enable them to enjoy more of the fruits of development (1996: 24).

And so it was that the economic backwardness of the Malays, and the need to address and resolve it became the center of government concern post-1969, leading eventually to the formulation of the New Economic Policy (1970–1990)—a twenty year policy aimed at eradicating economic inequities between Malays and non-Malays via the provision of an active national affirmative action policy whose aim it was to increase Malay and Bumiputera representation in all arenas of public engagement like education, the civil service, industry, commerce and ownership of share capital.[5] The presumption here was that since social and economic disparities between the Malay and Chinese communities were responsible for the tensions leading up to the May 13 riots, such a re-distributive policy could aid in ensuring against, or in preventing future outbreaks that may threaten the nation's well being.

But this story that I tell here is a story that cannot even begin to capture the meanings and implications such constructions have had on the psyche of the nation and its communities, capturing merely the process of its constitution. Drafted to provide a [economic] level playing field for Malays in an attempt to rectify the uneven nature of the nation's history, the NEP also created in the process of constitution, a privileged and a non-privileged group—Bumis and non-Bumis—and institutionalized as part of a deliberate policy, virtually all actions thereon. This contributed, in turn, to the inadvertent construction of a barrier and a structure that forevermore separates out the "insiders" and the "outsiders": in effect, serving in this instance as an example of nation as simultaneous terrain of inclusion and exclusion (Chatterjee, 1993).

* * *

> *As I was to discover during my return to this nation of my birth, all else told, every single Malaysian today who has gone through this experience has a memory and a story of that day and those events leading to and coming after it. Regardless of their proximity to the center of events, all are equally valid and real representations [partial though these may be] of the turning point in the young nation's history; maybe more significant in the final analysis because this partialness constitutes the real experiences of people as these are lived on an everyday level, not to mention the institutional and policy implications that were to come on the heels of this watershed event.*
>
> *" . . . This must never happen again . . ." and " . . . we didn't think we could kill each other . . ." are refrains all Malaysians hear in one way or another to this day. Refrains that I heard in multiple forms*

> *from multiple sources in multiple languages—a testament to the memory of the time when it did happen, and we did kill. Scars in the national psyche that remain all too real and must be acknowledged at the very least, for partially constituting the subsequent and current strategies of the government to "even" the playing field, to "consolidate" and articulate a common ground, to "unify," to search for "consensus" and so on: in short, to nation. For at the center of, and informing such efforts is a concerted attempt to come to terms and deal with a polity who is neither homogeneous nor completely united; themselves a partial legacy of the nation's heritage as a former European colony—a heritage that the nation still has to deal with and negotiate through today.*

* * *

However, as noted variously by observers of Malaysian society, the policy in the very act of its passage, created for the Malay and Bumiputera communities, a conundrum, an identity and a constitution as problems to be solved—as being never good enough. It is a constitution and a "national sacrifice"[6] from which they must constantly and forevermore negotiate in subsequent dealings and interactions with the larger Malaysian society. To be forever burdened, it seems, by the identity as individuals capable of moving forward and succeeding only with the help, the active intervention and/or the largesse of the state, regardless of capability and potential.

That this consequence is one that is not entirely unexpected is also true, evidenced by Mahathir Mohamad's (1982) implicit acknowledgement about the potentially "crippling" effect such a constitution or policy of affirmation might have on the community. As he notes in his argument for such a policy:

> . . . preferential treatment for Malays [ref. education] . . . are not a manifestation of racial inequality. They are a means for breaking down the superior position of non-Malays. . . . The Malays are not proud of this treatment. They are not proud of this "privilege" of being protected by law like cripples. They would like to get rid of these laws if they can, but they have to let pride take second place to the facts of life (Mohamad, 1982: 76).

Thus, the "facts of life"—namely Malays' economic inferiority in the face of non-Malays' superiority—would override pride in this battle to constitute a more even playing field for the community. What had been intended as a response to the community's lack of voice also in this instance had the effect of *silencing voice* within this same community. It is no small irony that in so doing, the future prime minister inadvertently contributed to the constitution of a backward and crippled Malay community for whom he is speaking here. It is within this historicized and contested backdrop, and in answer to such concerns and consequences, whether intended or unintended, that recent adoption of neoliberal policy by the state acquires its meaning.

* * *

Boundaries drawn that are two-edged swords—a national sacrifice, a microcosm in many ways of our "lack" of belief in the self [as former colonies, as underdeveloped], and a legacy of the ways knowledge from the West has come to constitute us, of how our history is also the history of the West. Along with the denunciation such knowledge would demand of our own knowledge and way of living and being: looking always, or being forced to look always "outside" for our validation and existence. And while it is true that resistance, mediation and appropriation are also an endemic feature of this relation between the Rest and the West; even here, at times, it seems like we must constantly forge ourselves against this Other and this Outside. Where and how then can we draw the boundaries? What was destroyed in the process of "independence'? What was gained?

* * *

Before proceeding, it is important to note at this juncture that the implicit focus on representation and voice as exemplified above highlights the problematic issue of whose voice and whose knowledge it is that we hear when attending to what or how a nation or a culture "is." The presumed equality of representation associated with *all members* of a community, culture or nation which underlie most mainstream cross-cultural management literature gets partially called into question by instances such as these where some, it would seem, have more right to speak than others. Articulated, as an example, by calls to collaborate with local researchers in order to effect greater cross-cultural responsiveness (e.g., Boyacigiller and Adler, 1991), this presumed equality seldom holds up for question the issue of who exactly it is we collaborate with or listen to.

NEOLIBERALISM AND THE NDP: RECONSTITUTING THE BASIS OF [MALAY] BACKWARDNESS AND NATIONHOOD

According to the story thus far, rather than a homogeneous terrain, Malaysia-the-nation is characterized instead by a heterogeneity and plurality in its social, economic and historical make-up. Indeed, as we see from struggles to articulate the inequalities and frustrations which is the nation's history, it is precisely this heterogeneity and plurality that have been and are underwriting the specific constructions and actions of the state, not to mention contestations within and without. There is neither a natural homogeneity nor an endemic equality that can be assumed in connection to it as a nation.

By contrast, as I illustrate in the segments below, recent appropriations of neoliberal discourses in the service of national unity indicate instead that Malaysia is an entity that is constantly in the process of coming into being: a process constituted simultaneously by local and global intersections of one kind or another, whether epistemic, political, economic and/or social. More specifically, certain features of the discourse are actively appropriated by state planners (1) to transition *away* from the NEP's essentialized focus on race, (2) to deal with the fragmentation induced by its national affirmative policy, and last but not least, (3) to confront the continuing concern with development in the nation, whether Malay or non-Malay. The latter, in particular, is closely connected to what the government sees as a continuing dominance of Third World economic independence by certain Western institutions of rule.

While NEP policies did contribute to the urbanization and increased representation of Malays in certain economic sectors and in ownership of share capital ['official" indices of poverty eradication and progress under the NEP], the failure of various programs aimed at increasing Malay participation on the entrepreneurial and management front remained a concern (Mohamad, 1993a). Of even greater concern to Mahathir Mohamad, affirmative policies under the NEP had led to the unintended consequence of providing "too much protection" to Malays, cultivating a dependency through its multiple forms of assistance that had neither led the community out of backwardness nor generated the independent self-sufficiency capable of inducing such a transition. To quote him: "the indiscriminate distribution of wealth which is immediately frittered away does not only defeat the objectives of poverty eradication or restructuring, but they do lasting damage by creating a very dependent society which cannot manage without continuous government support" (Mohamad, 1993a: 431). The problem, in light of such conundrums, is how the state might more effectively confront the continuing problem of developing the community.

While the earlier review on the backdrop to the local terrain had focused on a multitude of global factors and externalities—the worldwide recession, falling commodity prices, declining government expenditure, and multilateral development policies—which had compelled the adoption of neoliberal policy by state planners, what we see here and below are concerns specific to the nation and its history that are propelling the state to transition away from the NEP's policy of constructive protection. Hence, the state's "own" interests in, and active appropriation of neoliberal policy as a local alternative that does not rely solely on state initiative to effect the desired changes toward Malay development and national prosperity.

Under the market-mediated auspices of neoliberal discourse, what is changed is the very constitution of the problem itself. While the Malay community is still a problem to be solved, the community is seen less as essentialized possessors of certain racialized characteristics, but as lacking that variety of skills, knowledge and experiences essential for success in a modern society.

Central to this change in problem identification is the active reconstitution of the Malay community and citizenry into *economized entities* such as human resources, producers and labor. Throughout the prime minister's introduction to Vision 2020 (Mohamad, 1993b), members of the community and the citizenry are addressed less by their location as members of communities (e.g., Malay, Chinese) than by their location as "human resources" (: 415) or owners of particular "talents," "skills," "abilities" and "know how" (: 415). As he notes:

> [i]n our drive to move vigorously ahead nothing is more important than the development of human resources. . . . [i]t is blindingly clear that the most important resource of any nation must be the talents, skills, creativity and will of its people. . . . [o]ur people is our ultimate resource (: 415).

Constructed under the discourse of human resource development, at the center of the NDP is an explicit acceptance of neoliberal prescriptions calling for investments in people and the acquisition of certain know how associated with their ability to participate in modern sectors of the economy like research and design, industry and entrepreneurship.

Accordingly, this focus on skills and resources as national strategy is important since the possession of knowledge and a well-trained polity are not only the "basis of power but also prosperity" (Mohamad, 1993b: 418). Through such "equations," the social polity is *commodified* into a form of resource, an inventory and a cache that must be invested in and developed. It is a constitution that recalls only too well Kiggundu's (1989) constitution of nations as a composite of "assets" and "human resources" whose value is directly connected to their effective development and utilization. In this instance, it is a reconstitution which casts Malays in particular, and the citizenry in general, into new relations with respect to the state and the market: namely, as economic subjects, factors of production, labor and owner of technical or technocratic know how whose identity and value are derived primarily *in relation to* the market and the modern economy.

Put another way, Malay backwardness and solutions to it acquire meaning now primarily through their value as producers and participants in the industrialization process. As noted by Mahathir Mohamad:

> There is a need to ensure the creation of an economically resilient and fully competitive Bumiputera community so as to be at par with the non-Bumiputera community. There is a need for a mental revolution and cultural transformation (1993b: 407).

While alluded to in his writings on the Malay dilemma, the cultivation of "economic resiliency" and "competitiveness" take on greater impetus under the NDP. Relying still on the Bumi-nonBumi opposition, the inferior location of the former is used in this instance, to call for greater resiliency and competitiveness as opposed to constructive protection.

As such, the focus is not just on the acquisition of specific know how deemed necessary to be modern, but on the acquisition of very particular values needed or associated with the economized or organization-centered location of modern market subjects. I refer here to that coterie of features, values and so on associated with and expected of modern industrial and organizational subjects noted in my earlier review of the international management literature. Accordingly, it is their "achievement motivation," "devotion to know how," "self-improvement," "work attitudes and discipline," "managerial abilities" and "attitude towards excellence" (Mohamad, 1993b: 415) which counts. Hence, the focus on and need for a "mental revolution" and "cultural transformation." In short, in this era of the NDP, industrial discipline and the "development of appropriate economic cultures" (Mohamad, 1993b: 407) will be the barometers by which people and activities are judged.

Rather than simply attending to their inherent inferiority as a social community whose well-being can only be advanced with the aid of the state, this reconstitution of Malays into economized entities served not only to bring them into the modern sphere of the market, it also provided the state with the discursive entry-point to *shift policy attention away* from the state-support encoded by the NEP. "Self-improvement" and "economic resiliency" will now be substitutes for state intervention. What is at stake here is not just the issue of Malay backwardness, but reforming and reorganizing the government as well:

> . . . . If they [Bumis] are not brought into the mainstream, if their potentials are not fully developed, if they are allowed to be a millstone around the national neck, then our progress is going to be retarded by that much. No nation can achieve full progress with only half of its human resources harnessed. What may be considered a burden now can, with the correct attitude and management, be the force that lightens our burden and hasten

> our progress. The Bumiputeras must play their part fully in the achievement of the national goal (Mohamad, 1993b: 415–416).

What we see above is a reconstitution of the Malay community from poor and isolated individuals to un-harnessed resource, as well as a reconstitution of the *role of the state.* What was a crucial component of the national program of "constructive protection" (Mohamad, 1982: 31) under the NEP is a potential "millstone around the national neck" and source of "retardation" and "burden" under the neoliberal auspices of the NDP. By extension, greater impetus is placed on the community to "help itself." It must, in other words, "play its part fully," or as noted elsewhere: "[m]uch of the work of pulling ourselves up by our bootstraps must be done [by] ourselves" (Mohamad, 1993b: 407).

Thus the neoliberal reconstitution of Malays and the NEP is also directed towards the governing self and industry. According to the prime minister: "there will be a more selective approach which takes into account the need for quality and sustainable result" (Mohamad, 1993a: 431). Under this re-orientation, state regulation or involvement is systematically recast as a form of dependency and over-protection. Or as Mahathir Mohamad (1993b) notes in reference to private sector [de]-regulation:

> . . . productive liberalization ensures that our private sector will be less reliant on artificial profits and on protection. . . . Infants must grow up. They must grow up to be sturdy and strong. And this cannot be done if they are over-protected (: 414).

What is being organized here is the "productive liberation" of industry. While Malay dependence was constituted as a burden and retardation, industry dependency is constituted here as a form of "infant[ilism]." Constructed under these auspices of maturity and independence, this easing-away of the strong-hand of the state is part of a "national efficiency strategy" geared at "enhancing [the] competitiveness, efficiency and productivity in the economy" (Mohamad, 1993b: 411) as well as "reducing the administrative and financial burdens on the government" (: 411). As noted by numerous critics (Sachs, 1992) in their deconstruction of tenets central to the development sciences, these are all features of subjects necessary for the transition of nations to market-based economy.

In sum, what is being constructed in the foregoing passages is the active production of a "new Malay" [or "Malaysian" as the case may be], and a new basis for growth: economically resilient, fully competitive, self sufficient, independent, disciplined and in possession of those knowledge or productive and innovative capacity deemed necessary for development itself. This, in turn, is a possibility that relies on the reconstitution of the community and citizenry into economized entities. Targets of a like process of construction, local industries and government agencies are being reconfigured into facsimiles of modern market-mediated institutions: fully productive and efficient, and able and capable of providing the nation with the support necessary to compete in the global arena.

For the Malay community in particular, the transition called for is no less than a movement away from the implied dependency of the NEP towards the independent self-sufficiency of the neoliberal NDP, from being clients of the state to being fully participating citizens of the economy. In the process, this reconstitution to identity and subjectivity, backwardness and development brings to visibility important new features regarding Malay backwardness such as their lack of required education and skills, which *unlike* the environment or heredity, can be responded to and dealt with.

Concurrently, this appropriation of and focus on subjectivities and values connected to neoliberal discourse also provides the government with an alternative basis for *reconstituting relations between* the nation's plural communities. Under the NDP, great emphasis is placed on the development of a *cultural body* and a *cultural identity* that is singularly *Malaysian* in its emphasis on culture, religion, family and so on as the backbone founding such economic and structural changes to come. Unlike the exclusionist appeals of the NEP with its implicit and explicit privileging of the Malay community *over* the other communities, the NDP is an appeal to the citizenry at large, and a policy declaration to members of the private sector—where non-Malays constitute a significant presence—to participate in and forward this "vision" of the nation.

Differences in economic well being will no longer be primarily identified with race or ethnic community but with the type of knowledge acquired, type of economic participation, individual productivity or initiative and so on. Under Vision 2020, economics and economic forms of categorization, such as productive-nonproductive, professional-nonprofessional, entrepreneur-non-entrepreneur will be key identifiers for locating citizens, regardless of race, religion and so forth. In other words, it is *their link or contribution* to the economic development of the nation that counts:

> . . . the [NDP] policy should encourage the mobilization of all our resources and the utilization of the creative potentials of our *multiracial* society to build a strong economy and make the country more resilient against the instabilities and uncertainties of the world economy. We believe that each race has its own character and strength that can complement the others. The NDP, therefore, has been formulated to provide opportunities for *all Malaysians* to play their part more effectively in the development of the country (Mohamad, 1993a: 432, emphasis mine).

We see in the quote above a utilitarian focus on "mobilizing resources," "utilizing creative potentials" and so on. Framed within the fragmented and heterogeneous backdrop of the nation, the government's focus on economics as terrain of engagement and social identification is nothing less than an alternative away from its previously divisive reliance on race as a basis for state policy. This attempt to circumnavigate state-produced racial-fragmentation in favor of economics as a basis of community and identification is well exemplified by the following statement from the prime minister:

> . . . [since] we can't remove the racial (vertical) divisions easily, so we must try to remove the social (horizontal) divisions: then there will be poor Malays and rich Malays and poor Chinese and rich Chinese . . . once you come to that, there may be class division rather than racial division! [Mahathir Mohamad, as quoted by Khoo (1995: 125)].

As such, what we see also encoded in these reconstitutions to the nation's pluralistic terrain is an active focus on and acknowledgement over the "potential" which is its multiracial heritage, not to mention the positive contributions or "strengths" which each race or community can bring to aid in the nation's bid for development. Thus, what was once a divisive focus centered on "aggressive immigrant races," "Malay dilemma," "Chinese domination" and so forth is now "multiracial heritage."

This change in how nation and polity are constituted is underlined by the emphasis on *all* Malaysians and the respective parts that all individuals can play in the service of the nation's interest. In the following, constant reference to signifiers aimed at creating a sense of unity is emphasized by the prime minister:

> There can be no fully developed Malaysia until we have finally overcome . . . the challenge [the first of nine listed] of establishing a united Malaysian nation with a sense of common and shared destiny. This must be a nation at peace with itself, territorially and ethnically integrated, living in harmony and full and fair partnership, made up of one *Bangsa Malaysia* (Mohamad, 1993b: 404).

In this way is a *Bangsa Malaysia* (Mohamad, 1993: 404) as opposed to a *Bangsa Melayu* to be had [In Malay, "Bangsa" can be used for referring to race, nationality and/or descent; "Melayu" refers to Malay]. What is being produced here is a shift in the way that the state constitutes itself as a national and cultural entity: in effect, moving from a focus on race to citizenry or nationality as its barometer of nationhood.

Thus, when looked at from the social and historical terrain of Malaysia-the-nation, Vision 2020 and the NDP is not just a simple series of prescriptions on how the nation may transform itself into a developed entity. It is also an explicit *social and political strategy* aimed at developing and uniting the nation, albeit via the progressive economic homogenization of an otherwise heterogeneous and socially demarcated polity. In particular, the economized features of the discourse are actively used to construct a common ground, deemed to be capable of crossing boundaries or navigating the differences in race, class and culture locating the nation's communities.

It needs noting at this juncture that while such changes in the constitution of the national polity and the actual development activities resulting from the NDP have certainly been criticized in terms of its content, process and/or effects (e.g., Kahn and Loh, 1992), to imply that such a nationalized constitution of communities—economized and westernized though it be—is one that is totally unwelcome is also not true. As I was to realize during my time in the field, there is also much support across the board from Malaysians of all ethnic communities. For embedded within this national vision, particularly for the Chinese community, is a memory of the exclusion and the silence that is also the nation's history. This constitution of a more "inclusive" notion of nationhood and citizenry, rife though it is with potentially new and increasing inequalities, is preferable in multiple ways, for multiple reasons to the past alternative for certain segments of the Malay and non-Malay communities. Although it needs noting by way of a caveat that this preference does not necessarily imply "acceptance" of modernist assumptions regarding the viability of either markets or democracies. Rather, this preference is one directly framed by, and against, the history and memory of communal relations specific to the nation.

In sum, we see in the explication of Vision 2020 policies, an appropriation of neoliberal discourses to the exigencies of the local situation. On the one hand, it is actively appropriated in response to the nation's continued concern over the problem of Malay backwardness, and on another, actively used to provide an alternative basis for reconstituting nationhood itself. Both rely on an active constitution of the subject that is implicit in the discourse; namely, economized entities and resources whose value is derived by their connections to the marketplace and economy.

VISION 2020 AS "ECONOMIC DEFENSE" STRATEGY

Also embedded in this policy transition towards a neoliberal bent is an effort to defend both nation and economy against the continuing domination of developing economies by a coterie of Western, industrialized nations. The adoption of market-oriented and structural-adjustment policies such as adherence to free trade, expansion of the domestic market, focus on exports, and greater government-business collaborations, along with active attempts to "influence the course of international trade" and "oppose the formation of trading blocs" are ways for the nation to increase its "economic defense capabilities" and "marshal influence" in the international economic arena (Mohamad, 1993b: 418). According to the prime minister, "[w]e must . . . play our part and not passively accept the dictates of those powerful nations who may not even notice what their decisions have done to us" (Mohamad, 1993b: 418).

As observed by Khoo (1995) in his assessment of Mahathir Mohamad's critique of north-south relations, at the center of the prime minister's critique is the existence of an industrialized and economically powerful West whose economic well being has come on the backs of most developing economies. In particular, their reliance on developing nations as a source for primary commodities and market for manufactured goods, their control over terms of trade, technology and "willful manipulation" of commodity prices and exchange rates have threatened to overwhelm the very economic foundations of most Third World nations. The collapse of the Malaysian economy during the mid-1980s resulting from the general collapse of commodity prices and falling investments seemed only to confirm this scenario of inequality and dependency. The prime minister sees in the actions of this wealthy and powerful elite:

> . . . free traders of convenience . . . [who] fanatically proclaim themselves to be the standard bearers of free trade except when it affects themselves adversely (Khoo, 1995: 59, quoting Mahathir Mohamad).

Or, as noted elsewhere in his call to local industry and nation to engage in greater international trade, and more active participation in a variety of regional trade organizations:

> . . . there is no such thing as free trade. Trade was free when we were the markets for the rich developed countries. But when we want to gain access to the market of rich countries to sell our manufactured goods, restrictions are imposed. The very people who taught us to abide by the rules of competition are now violating these same rules for their own political and economic reasons (Mohamad, 1993a: 442–443).

Accordingly, efforts such as the formation or participation in trade blocs are necessary in order to "*counteract* some of the negative effects of rising trade protectionism by certain developed countries" (Mohamad, 1993a: 443, emphasis mine).

In contrast to the oft-constructed image of free market economies forwarded in the West which sees such open economies as arbiters of equality and the free movement or exchange of goods and resources (e.g., Edwards, 1995; Jorgensen, Hafsi and Kiggundu, 1986; Kiggundu, 1989; Lieberman, 1990; Vernon-Wortzel and Wortzel, 1988; World Bank, 1993a), what is forwarded here is a commentary about the *unequal* feature of international trade and exchange, whose process and consequence are derived less from the laws of supply and demand than they are from the political and economic self interests of a wealthy and elite few. When looked at from this perspective of the global economy, the counter-representations forwarded by Mahathir Mohamad actually *contradict* received presumptions about its universality and progressive evolvement towards an egalitarian world without borders [i.e., globalization]. As exemplified by his constitution of why adoption of recent economic strategies is important (e.g., the non-existence of free trade), neoliberalism is appropriated as an *active defense against* neoliberal policy itself. In effect, it is seen as a way of re-configuring relations of inequality between the West and the Rest, one that is equally capable of sustaining the self *against* this same West.

Thus, the "mobilization" and "utilization" of the nation's "creative potential" (Mohamad, 1993a: 432, emphasis mine) quoted previously is needed not only to develop or unify the citizenry, it is also required to "make the country more resilient *against* the instabilities and uncertainties of the world economy." Added to this instability and uncertainty is the necessity to defend against a political economy which pits developing nations against other like nations:

> . . . [a] major consideration in pursuing the socioeconomic goals of the NDP is that it is necessary to take into account the needs of the economy to compete efficiently in the international marketplace and to face the new challenges that are already emerging. For instance, all countries including the centrally planned countries are now liberalizing their economies and competing strongly for foreign capital. In the face of this challenge, Malaysia must enhance its attractiveness for investors and businessmen to expand their activities in this country (Mohamad, 1993a: 432).

Rather than a statement about the viability of neoliberalism, the above is a commentary about the *pressures*—global, institutional and economic—compelling nations like Malaysia to adopt such policies. As exemplified by the above, the "necessity to compete" [institutionalized via the investment strategies of multinationals, and structural-adjustment conditions of multilateral agencies] is a key reason why these nations take on the actions that they do.

All constitutions discussed above serve as examples of a *non-passive engagement* in those arenas of activity where the nation's defense capabilities are currently most vulnerable. To Mahathir Mohamad, the key to survival and success—given this continued domination of the Rest by the West—lies *precisely* in the acquisition of those elements associated with the economic dominance of the industrialized bloc that he so vocally opposes: technological and scientific proficiency, a viable local entrepreneurial community, industrial linkages, management know how and so on. *However paradoxical, it is a strategy aimed at defending the nation against western economic dominance by means of a westernizing economic defense.*

This paradox of simultaneous appropriation and constitution is an instance of how nations like Malaysia are constantly negotiating between silence and voice, subjectivity and agency; a condition connected to the power of modern discourses to constitute the visibilities of even those individuals or nations *not unaware* of the uneven consequences attached to these same discourses [and practices]. An instance, as it were, of how actions and decisions, however critical, can only ever be articulated within the "modernist rubric" (Manzo, 1991) organizing the discourses so long as assumptions and evaluation criteria endemic to the modern episteme are accepted as barometers for judgment and decision.

That this circularity is not totally unrecognized is evinced by the careful, albeit *still* paradoxical, emphasis placed by Mahathir Mohamad on how the adoption of such policies are neither undertaken blindly nor purely by state planners: it is "economic imperatives" not "politics and ideology" (1993b: 418) that underpins the course the nation is setting on. Not only are such models appropriated strictly for economic and one might add, defensive reasons, the *lack* of values and morality encoded in them is a primary reason why wealthy industrialized nations have been able to continue dominating the Third World. Malaysia should, accordingly, be developed "without being a duplicate" of these nations (Mohamad, 1993b: 403).

With this in mind, equal emphasis is placed on how the nation's attempt to industrialize must be constantly juxtaposed and framed within its larger cultural and social backdrop: "the comprehensive development towards the developed society that we want—however each of us may wish to define it—cannot mean material and economic advancement only" (Mohamad, 1993b: 406). In particular, the private sector, under the rubric of "Malaysia Inc.," has an important role to play here in terms of helping to develop the nation's human resources:

> Train your own manpower [sic]. Equip them for their changing tasks. Look after their interests. Upgrade their skills. Manage them well. And reward them for their contribution (: 419–420).

This focus on human resource development and its significance in connection to the Bumiputera community and local industry is particularly important because the private sector has a *social,* not just an economic responsibility.

Constructed under a narrative of self-interest and collaboration, "the private sector will be called upon to share a larger burden of training the human resources of the country . . . to enable the output from the training institutions to become more relevant to the actual needs of the private sector" (Mohamad, 1993a: 445). What is constituted above is not just a call for greater government and business collaboration, but an active inclusion and drawing in of the business community as *part of* the state's bid to modernize in general, and to aid the Bumiputera community in particular. In a special aside to the private sector during his speech to the Malaysian Business Council, the prime minister reminds the community that the nation's:

> . . . struggle to ensure social justice—to uplift the position and competitiveness of the Bumiputeras and to achieve the other social objectives—must be your struggle too. . . . Malaysia cannot deregulate if bankers eventually behave as "banksters," if the freedom afforded to enterprise becomes merely the freedom to exploit without any sense of social responsibility . . . (Mohamad, 1993b: 119).

SUMMARY

What has changed in the story told here is the boundary between a Self and an Other against whom the state has been articulating its policies. Specifically, there is a change in terms of what the problem of Malay backwardness is, and how it should be resolved. Under the NEP, this backwardness was attributed to a variety of characteristics endemic to the environment and heredity. Chinese economic dominance was constructed as a problem and an Other against which calls for constructive protection by the state was articulated. By contrast, under the NDP, the primary problem lies in the lack of certain elements—education, technology and discipline—deemed necessary for development. Economic dominance by an industrialized West and the political economy organizing relations between this same West and nations like Malaysia is now the Other against which activities of the state are defensively forged.

Neither under the NEP nor the NDP is there a questioning of the concept of backwardness or underdevelopment itself. Changes in policy were more a matter of prescription than form; an "internal" discussion about what is needed to resolve the overarching question of underdevelopment and equality. To this end, the simultaneous appropriation and application of western development strategy and practices informing Mahathir

Mohamad's modernization activities contain within it certain crucial paradoxes. It is one of the ironies of the story that the very adoption of such a westernizing economic strategy as embedded in the prescriptions of Vision 2020 is a form of engagement steeped in those very values whose outcomes the prime minister criticizes.

On the other hand, while participation in the global economy is an implicit acknowledgement of the nation's intent to play in the international (economic) arena, the meanings attached to the playing of this game are certainly very local ones. Global economic and political compellers notwithstanding, this decision to "embed" the self in a global language and terrain is in multiple ways, a statement and a symbol of "having arrived," a beating of the West and former colonial powers at their own game, and of "winning in spite of the odds." As I was to realize, such sentiments are to be found not just in comments from officials, but also from people I met in unofficial capacities or circumstances. As noted by a *towkay* [local proprietor, usually Chinese] in my conversation with him about the nation's economic performance and increasing prominence in the international arena, this visibility is something to be proud of [*we will show them . . . see if they (i.e., the West) will look down* on *us anymore*]. Spoken in Cantonese, "show" literally means show as in a performance.

Industrialization in the nation is thus *something less and something more* than economic competitiveness and development. Embedded within is the symbolic currency represented by, and conferred on Vision 2020 as an index of Malaysia's entry into the global economic arena. A reconstitution of history, the government's development vision is actively used to unite and inculcate a sense of [national] pride amongst the polity.

What is significant from the standpoint of cross-cultural knowledge is how meanings attached to the changing boundaries of nationhood, and the nation's development impetus—what may on first encounter seem transparently western and modern in nature and feature—can be or are distinctly local[ized]. What is an ostensibly western form of organization and institution is adopted and appropriated here to fit the exigencies of the local situation. Neither is this just an issue of adaptation or refinement; implicit in this localization is an active appropriation and transformation of current development models' economic feature into a social and political strategy. Concurrently, support for this strategy exists not only because of some memory of past exclusions, but because it is an index of a not-quite-dominant community finally coming into its own. Thus, development as arbiter of nationalism is by itself transformed from the singularly modern narrative of evolution and progress to avenue and symbol of resistance. In so doing, what is changed is the specific constitution of relations between development and nation-state. Both are constituted instead as barometers *against* a history and a modern West.

In sum, this chapter has been aimed at providing an overview to the context surrounding Malaysia's adoption of market-mediated models of development. Through its historicized and discursive review, it pinpoints attention to the dynamic interconnections between culture, history, economy and knowledge that are constantly organizing the nature and forms of how the local is *intersecting* with the global. Implicit within the approach is a commentary about the difficulties of understanding or reading Others *in the absence* of this historicized and discursive frame.

In Malaysia's instance, little can be known about "the local" unless attention is given to the historicized and cultured conjuncture between knowledge and power, be it in reference to Malay-Chinese relations, the symbolic currency of Vision 2020, reasons underlying the adoption of specific development policies or choices, meanings attached to such choices, the government's critique of an economically dominant West and so on. Moreover, by focusing on changes occurring in the nation as an interpenetration of locally driven concerns and globally infused language, the chapter highlights the simultaneous process of constitution and appropriation that are at play: a simultaneity that calls into question the location of "nation" as barometer of one culture, one people and one goal; and "development" as an universal and global trajectory of epistemic, social and economic possibility.

From the standpoint of institutional ethnography, such an account and the concerted interconnecting between global factors and local concerns embedded in it provide a way of situating the social particularities operating on/in the nation. In doing so, the methodology contributes to the possibility of generating a more locally attentive way of understanding what is framing and organizing local planners' interests in knowledge transfers to begin with. In effect, this situated and historicized ethnography is a way of rendering apparent those relations and interconnections which are implicitly and explicitly organizing actions and intentions at the local level; relations and connections which are otherwise silent or "not fully apparent" (Smith, 1987: 92) in this story of knowledge transfers.

Chapter Five
Quality in the Service of Development

> We cannot but aspire to the highest standards with regard to the skills of our people, to their devotion to know how and knowledge upgrading and self-improvement, to their language competence, to their work attitudes and discipline, to their managerial abilities, to their achievement motivation, their attitude towards excellence and to the fostering of the entrepreneurial spirit.[1]

As I traversed through a Malaysia seemingly bent on quality-controlling itself to the ninth degree, the question "why this particular interest in quality?" kept circulating through my head. It was not until I started configuring its localized interest in neoliberal policy that I was able to properly situate this focus in terms of the nation's development objectives.

Under the specific constitution of the problem engendered by the market-oriented gaze of Vision 2020, objects of development produced by the discourse such as Malays and the civil service are primarily resources whose value lies in their capacity to reconstitute themselves into—or *be* reconstituted as—valuable development entities. As a concept and a practice associated with institutions' ability to increase efficiency, productivity and competitiveness, "quality" is the cornerstone around which current efforts to educate and cultivate a modern subject are organized. Put another way, it is in response to the state's constitution of underdevelopment as a problem associated with the lack of an economically relevant subject, and the systematic construction of this same subject as too dependent and non-disciplined that *interest in* quality management and practices acquire its meaning.

Long a component of many multinationals' strategic endeavors and everyday activities, quality know how is actively appropriated by state planners, not only for what they can teach workers as a specific site of engagement and action, but also to explain and locate current efforts undertaken by the state in the name of development. Included here are the state's restructuring or corporatization efforts, human resource training activities, and business-government collaborations. In effect, quality is a key condition of discursive possibility enabling state planners to explain and forward their human resource and development objectives.

Seen as a barometer of efficiency and effectiveness, quality management and practices are used by officials and managers alike as a way of conveying the importance of modern education, training and industrial discipline. As noted by a senior official from the National Productivity Corporation (NPC), whose motto is "Productivity and Quality": "a quality workforce is what will make a big difference for us in the future"; or as forwarded by the head of the Malaysian civil service regarding changes being introduced to improve the "business of public administration":

> These changes represent a fundamental shift in paradigm. . . . Values such as *quality, productivity, innovativeness, integrity, discipline, accountability* and *professionalism* underpin the efforts to institutionalize a culture of excellence. . . . *Total Quality Management* . . . help[s] civil servants discharge their duties in an efficient, effective, responsible and responsive manner. . . . training programs have been designed to equip civil servants with not only the necessary knowledge and skills but also the essential values and mental orientation to perform their duties effectively (Sarji Abdul Hamid, 1994: vii).

Used to explain the shift in paradigm, total quality management (TQM) is a *site* for acquiring knowledge and skills, as well as for enculturating values and mental orientations deemed necessary for having a culture of excellence. Framed as it is under the auspices of Vision 2020, it is no surprise that these words recall only too clearly Mahathir Mohamad's (1993b) call for "industrial discipline" and "mental revolution."

This chapter provides a picture of what was put into place in the service of the nation's aim. Focusing specifically on quality management and practices, it starts by looking at features of the quality discourse being actively appropriated by officials in their bid to constitute for the nation a disciplined and independent subject, able and capable of showing the initiative and self-sufficiency required by both nation and market. This is followed by a look at the specific forms of quality know how being deployed to locals by managers and trainers in the multinationals and statutory agencies I had visited, and their significance in light of the aforementioned aim. The chapter ends by looking at the nexus of institutions, practices and relations that were put in place to constitute, coordinate and deploy this targeted body of knowledge.

My intent in this chapter is to situate the phenomenon of and interest in quality within the social particularities in which it is embedded, both local and global. Specifically, by referring our attention to the myriad of assumptions, concerns and practices that are organizing this interest in quality, the chapter brings to visibility the constant presence of a West, whether epistemic, economic or institutional, who is centrally involved in [re]-constituting actions and intentions at the local level. In particular, it focuses attention to the active processes and/or procedures of transformation that this appropriation of quality know how entails: namely, the "what" and the "how" of knowledge transfer.

TOWARDS THE SELF-INQUIRING AND SELF-KNOWING WORKER

In this segment, I look specifically at some of the procedures of knowledge and power being put into effect to reconstitute locals' subjectivities and regulate their behavior, whether individual or institutional. Of immediate significance for understanding this story about the discursive and constitutive power of knowledge is the writing of Foucault (1979) on power as a "positive force." To quote him:

> We must cease once and for all to describe the effects of power in negative terms. . . . In fact power produces; it produces reality; it produces domains of objects and rituals of truth. The individual and the knowledge that may be gained of him belong to this production (Foucault, 1979: 194).

Or as alternatively noted by Aihwa Ong (1987) in her ethnography of neophyte Malay factory women whose lives and locations in their respective communities are being systematically transformed via the introduction of modern industrial discourses and factory discipline in Malaysia:

> Foucauldian concepts are directly relevant in delineating the modes of domination within high-technology factories . . . [because it] identifies and exposes the power effects of knowledge/power on individuals, their reconstitution [as subjects], and self-management. In social systems mediated by technology, such as transnational corporations, coercion is enforced not only through labor relations but also through corporate discursive regulation of work performance, comportment within and outside the factory premises, and the workers' sense of self (1987: 142–143).

As in Ong's study on how these women's sense of selves are being constituted and reconstituted in Malaysia, the constructions of locals I document below have to be situated within this process and these procedures of knowledge and power that I call attention to here. From the standpoint of neoliberal development, these constructions are procedures relied on by officials and managers to effect the transformation of locals seen as necessary for facilitating their integration into the global marketplace and economy.

At the center of interests in quality management and practice is a problematically youthful and undisciplined subject. Already framed within a reconstituted notion of subjects as human resources, locals are often constructed by officials and training personnel from statutory and government agencies as growing subjects, either still too young or not quite able to play in the grown-up terrain yet. As noted by a senior official from the National Productivity Corporation (NPC):

> . . . the workforce is young, we have to push people—especially the operators—to be more quality conscious, otherwise they won't care or attend to it.

While similar in their reference to subjects' youthfulness, earlier constructions forwarded on female factory workers observed by Ong (1987) arise from a gendered and Orientalist constitution which see their youthfulness as symbolic of their "malleability" and "trainability" for work in modern factories, whereas the youthfulness of locals being constructed here is forwarded as a "problem to be solved" since it is representative of their lack of knowledge and discipline. Either way, both these young women and now these young workers [or Malays], become and are effective objects and candidates for training and transformation.

The preponderance of this youthfulness, a construction also bringing to mind the definition of local industry as *infants,* is such that incentives are provided specifically to encourage local small and mid-size industries to partake in such training efforts. According to the NPC official: "this is how the government encourages participation in quality training programs . . . otherwise *they won't go*" (emphasis mine). Thus, the concern here is not only a youthful, but a *reluctant* subject who must be induced to participate via the provision of certain regulatory and incentive programs.

These constructions recall Frederick Taylor's (1911) construction of workers as soldiering-prone and inefficient entities who must be actively taught and given the knowledge so that they may be more efficient on the one hand, and induced to participate via a combination of disciplinary and economic forms on the other. Just as these constructions provided Taylor with the discursive entry-point needed to begin refashioning the modern industrial workplace, the constructions of locals as youthful and undisciplined provide to local officials their entry-point into the terrain of the nation.[2] This connection serve as a reminder that what is at stake within officials' constructions of problems is the active reconstitution of subjects into efficient and productive industrial workers.

Specific features of the quality discourse are appropriated by officials and trainers not only to explain their actions and intentions, but also to specify the boundaries and parameters of how such transformations should be effected. Of particular significance here is the currency conferred on the practice of continuous self-assessment and self-regulation called for by quality control concepts and tools: namely, its focus on continuous problem identification, cause-and-effect analysis, responsibility and accountability through teamwork, quality control and monitoring. Take as an example the following explanation forwarded by the aforementioned NPC official in his reference to who is winning the Prime Minister's Quality Award:

> . . . so far, only MNCs have won the award. . . . quite a lot of [local] SMIs do apply, but they're still not winning [pause], sometimes they ask me if I think they will win, I tell them [pause], "there can only be one winner," but *it is good for them,* it gives them an *opportunity* to assess [emphasis] their capacity and potential—*how they do* on quality and productivity—so I tell them, "take this opportunity to assess yourself."

It is self-assessment, and the opportunities for such assessments afforded by participation in TQM activities that are crucial. What we see embedded in this construction of subjects as "yet to be good enough to win" is a message on *how to win:* that is, to become True members of the modern elite is to become effective in one's capacity to evaluate and transform the self into an independent and self-sufficient *agent* or *site* for change. Moreover, change requires not only that attention be constantly focused *back* on the self; it requires that individuals are equally able and capable of seeing in this self, a problem in need of education and discipline.

Reminiscent of Foucault's (1980) notion about "the confessional"—a point observed by Townley (1994, 1993) in her deconstruction of modern human resource management practices—this focus on self-assessment is a form and procedure of knowing that simultaneously relies on the presumed *capacity* of the self to know the self. To the degree that this is so, current interest in and transfers of knowledge are directly connected to officials' and managers' needs to provide for workers, very specific *ways of knowing the* self.

As I seek to show in following segments, it is a knowing that is mediated by very specific conceptual, textual and documentary forms (e.g., quality control tools and schemata) whose intention it is to produce a very particular kind of subjectivity: economized, cooperative, functionalist, efficient, rational and productive. In short, that compendium of knowing and doing—indeed, of being—which are central to/for the existence and sustenance of modern workplaces and economy (e.g., Barnard, 1938; Fayol, 1949; March and Simon, 1957; Taylor, 1911).

As noted by one of the staff managers at Infotech in his explanation about the importance of a new strategic activity in the company geared at identifying how employees feel about their development as workers:

> . . . we try to find out what the needs of the people are . . . it is a way of *gauging yourself ah* . . . the individual by answering it assumes responsibility for his [sic] own development . . .

As with self-assessment, the object of attention here—a questionnaire with predefined answers pertaining to the adequacy of training, scope for development and so on—is intended as a way for workers to not only gauge themselves, but to assume responsibility for and over the self.

This reference to self-responsibility—and the implied capacity for "initiative"—is one that is already framed by discourses in the organizational behavior literature concerned with the need to specify for both managers and organizations alike, the exigencies, processes and/or features of what the modern worker should be like (e.g., Argyris, 1957; Barnard, 1938; Herzberg, 1966; Katz and Kahn, 1966; McGregor, 1960). Specifically, features and processes that include attentiveness to work values, motivation, commitment, skills or capacities and so on (e.g., Hofstede, 1987; Kanungo and Wright, 1983; Kiggundu, 1989; Putti, Aryee and Tan, 1989; Ricks, Toyne and Martinez, 1990; Torbion, 1985) required for effective and/or efficient performance.

This focus on the "how" and "what" of having organizationally and economically relevant subjects is organized in turn by the need to cultivate a very particular way of being and knowing needed to sustain the market as a functioning social system and institution. As noted by Berthoud in his deconstruction of the market as a concept and institution linked to the development of modern society:

> To conceive of the market system as a man-made institution rather than as a self-creating, self-perpetuating order is a way of recognizing that the market is controlled by various traditional political, social and moral constraints, and reinforced by a number of political and cultural innovations. In other words, *the existence and expansion of the market is dependent on institutions and cultural values* (1992: 79, emphasis mine).

Thus, interests in values and other specifications regarding what is needed to be a "productive" worker is already coming from this need to actively [*re*]-create and sustain the social conditions up-holding modern workplaces and economies.

In this sense, the call for industrial discipline and mental revolution fielded out by Mahathir Mohamad (1993b) and other local officials in Malaysia are very much involved in this process of producing subjects who are relevant and needed to sustain the transition of the nation to a market-based system. Put another way, it is in connection to this need that knowledge transfers and human resource development activities are crucial sites for effecting the transformation of local workers into modern organizational and market citizens. As Ong (1987: 3) observes, in industrializing nations like Malaysia, "continuous activities through education, media, and employment structures are required" to facilitate the integration of locals into the global economic arena.

Accordingly, there is a constant barrage of messages conveyed to people and workers as to what is considered desirable to have or to be in terms of their behavior, values and character *as* members of both nation and modern workplaces. As noted by one of the human resource managers from Infotech during a talk to prospective interns and employees at a local university, speaking here on how they might distinguish themselves from others: "you cannot behave like employee *biasalah* ['biasa" in Malay means normal], you have to take initiative*lah*."

What followed during much of this manager's talk is virtually a practical guide specifying what people should do and not do, and how they should be or not be as members of these modern workplaces. The following is an excerpt from his talk:

> . . . make sure you know the personnel, training managers . . . the department you need to go to, and people responsible in that area . . . some people don't even know after two weeks. . . . Once you know, you must get everything correct . . . you have to show your interest and seriousness from the first day—that's image *ya tak?* . . . *Kalau you first day itu hidup tak hidup, tidur tak tidur* . . . *susahlah!* [contextual translation: "if you're half dead or half asleep on the first day, then it's going to be tough going!"] . . . work out a time chart of your activities . . . identify people who can help you . . . follow through . . . [there is] no spoon-feeding. You have to work everything out yourself . . .

Implicit in this seemingly innocuous list is the systematic construction and representation of what is constituted as *the* desirable worker: namely, the prototypically self-sufficient [*there is no spoon-feeding, you have to work everything out yourself*], self-regulating [*make sure you know, work out a time chart of your activities, follow through*], and productive [*get everything correct, show your interest and seriousness*] employee able and capable of showing initiative. Anything other than that conveyed is, under such a constitution, potentially a problem.

In a like manner, the various deployments of quality precepts and practices are constantly involved in constructing for workers a message on what is important and valuable *to have*—or *to do*—in order to be relevant to both the organization and economy—indeed, the market. As an example, the metaphor of "the chain" used by Multitech to convey the key concepts and procedures of TQM to new hires during their orientation construct for these workers the importance of being responsible and accountable when it comes to their respective tasks and actions. The following is a transcribed excerpt from the company's orientation video [used in the early to mid 1990s]—presented here with the permission of the company—where this notion of the chain is introduced to workers:

> The Chain. It can't do its job unless all its links are the same quality. . . . Linked chains, for instance, can withstand great tensile forces, while others are designed for high rotation performance. But each link in every chain is as important as all the others which makes the chain assembled for interconnected processes and industrial production.

Used variously to convey the interdependency of work and production activity, the metaphor is used here to signify and constitute for workers, their respective locations as "links" in an organization chain, a productive chain, or in the instance of these multinationals, as links in a global economic production chain. So constituted, each and all workers forming part of this chain have an equal responsibility for ensuring that the chain can do its job. This capacity to pull-one's-own-weight is crucial since its *absence* would mean that the chain *could not* function *as* a chain.

What is significant here is how this constitution of workers as links is also implicitly constructing for its audience their relevance to the well being of the organization. In this instance, it is about workers' ability and capacity to *situate the self* as an effective—as opposed to a defective—link in this chain. Thus, constructions such as those forwarded here are not just about the simple transmission of information; they are important instances and sites for constituting the subjectivities of workers *as required* by the organization. Or as noted by Reed (1999: 33) in his overview of the dynamics of organizational control in late modernity: "TQM-type initiatives" may be seen "as another managerial strategy for internalizing surveillance and control to the extent that they subject workers and lower-/middle-managers to a form of organizational discipline in which the dictates of market rationality become unchallengeable."

Accordingly, the discourse stipulates for workers specifications on what is deemed to be effective and relevant. Constitutions like the foregoing provide to managers and trainers the conceptual entry-point(s) they need for elaborating on what organizations require from their workers. In the following excerpt, the implicit call for self-knowledge and responsibility is reinforced and extended via a focus on what this relevancy may entail:

> Quality Assurance and Quality Control are two links that gets special attention. . . . Link for link, a chain's quality is evidence of continuity. . . . Constant control of the individual production processes guarantees the high quality of the optoelectronic components. . . . The logistics of this procedure shows that it is a chain. With statistical process control, each link, each phase of production is monitored. Continuous observation like this identifies defects early and remedies the causes . . . if any irregularities should occur . . . the thorough documentation of production data and an analysis . . . quickly lead to effective solutions. . . . Only constant monitoring of the production processes enables us to maximize the yield. . . .

What is important is workers' capacity to continuously monitor production processes, identify defects, document production data and so on. Mediated by the concept of quality assurance and quality control, these conveyances forward to workers the centrality of their participation and responsibility to contribute towards monitoring processes. Given the interdependencies of a chain, both ensure that workers are continuously engaged in needed activities of observation and monitoring. Put another way, these requirements are important to the organization since it is only through workers' involvement and participation in such monitoring and problem-identification activities that it can effectively know what is going on or going wrong. Hence, the import accorded to self-assessment and self-regulation noted previously.

Moreover, the description also actively constructs for workers the connection between such participatory and monitoring capacity to the economic and productive well being of the company, referenced here by the allusion to continuity and maximization of yield. Mediated by the discourse and practice of quality, it is a connection that relies on the ability and capacity of workers to be both self-aware and accountable: in particular, aware of the self as a link in the chain of modern organization/market citizenship, and accountable for the performance of both self and organization/market. In the process, what get constituted are very specific connections linking workers to organizations and markets.

Linking Quality to the Workplace and Economy

An interesting introduction to the pervasiveness of this quality-mediated link between workers, organizations and markets came in the form of what I have come to think of as "name-tag imprints." In almost half of my initial conversations with key personnel from Infotech, individuals—often in answer to semi-structured requests for information about what they do—would reach almost automatically for the combination name tag and security pass hanging around their necks, fiddle through the multiple credit-card-sized tags attached to it, and read off word-for-word to me, the company's objectives and key initiatives. Somewhere between my asking them about what they do and their formulation of a response is *the company, its objectives* and *TQM*.

The connections identified and reconstructed here between worker, company and quality are conveyed early on to workers by company representatives. The following statement, articulated by a staff manager from Multitech in describing the centrality and prominence of TQM practices, was forwarded to workers during a new hires orientation in the company:

> . . . Multitech is not paying our salaries, the guy doing this is our customers . . . we exist because of our customers . . . every job is essential and important ah . . . and so the buck stops here—we have to take ownership if we can . . .

In this manager's reference to who is paying the new hires' salary is a reminder of their location as waged labor or paid employee; the reference to customer satisfaction, a conscious linking of labor participation to markets. Mediating between these links is the company and quality control, referenced here by the allusion to customer satisfaction. What is under construction is not just the connection between people and the company, but the specific responsibilities expected of them in their capacity *as* workers. Like the self-assessment and monitoring requirements noted previously, "taking ownership" is what *links* workers to the aims of organizations.

Nowhere is the construction of this quality-mediated link between worker, company and market more apparent than it is in the following excerpt from the company's orientation and training video:

> Over thirty years ago, the quality of industrial products came to be associated with their traditional countries of origin. German cars, Swiss watches, Paris fashions. Today, the sign of quality isn't the country of origin, it's the manufacturer who sets the guidelines. "Made by" is the phrase that tells us about the quality.

> "TQM" is what we call our corporate strategy of providing customers with flawless products. By specifically motivating all our employees in all countries at all levels, we're able to lend equal weight to the phrases "made in [country]" and "made by [company]."

Not only is there a valorization of quality precepts and what they can do, quality is constituted less as a national signature of industrial capacity and excellence than it is an *organizational* one. Underlined by the reference *away* from "made in" towards the emphasis on "made by," this recasting not only contributes to the construction of a global, *trans*national world inhabited by multinationals, it is also able to effectively draw a connection between such manufacturing or economic veracity and quality management practices. In particular, the juxtaposition between TQM as corporate strategy and the association of this strategy to both flawless products and employees construct for new hires and audience alike, the specific message that both TQM practices and employee participation are central to the company's ability to achieve its aims and gain or maintain a place in the pantheon of transnational success.

Concurrently, the last sentence connecting "all employees in all countries" to the company's ability to "lend equal weight" to the phrases "made in" and "made by" also construct for the audience, the existence of a seemingly *globalized membership,* referenced by the focus on "made by," while *simultaneously* acknowledging the implicit national superiority [i.e., "made in"] of the company's home-location, from wherein such membership in the technological or economic elite is founded. These constitutions recall Chatterjee's (1993) reference to modern nations as simultaneous terrains of inclusion and exclusion, albeit effected here within an *organizational* context.

More significant still, the conveyance of this information to new hires right upon their entry into the organization pinpoints the importance of events such as orientations and training as crucial *sites* for constituting the experiences, subjectivities and locations of workers. From the standpoint of institutional ethnography, what is being organized here is the active framing of their everyday reality and meaning-making processes as organizational members and citizens of developing nation—mediated in this instance, by the documentary-like medium and modernist constructions of the video.

From the standpoint of a postcolonial analysis, this active constitution of worker subjectivity and location is even more apparent in the segment of the video following immediately from the above:

> [Images shown on the screen] A panoramic view of the company's plant in Penang comes on. This pans out to encompass an image of the plant with a Malaysian Airline jet in the foreground, caught in the process of approaching and landing at the international airport located close to the free trade zone where the plant is situated.
>
> [Narration] Theories put into practice. Chips produced in [Western country] are assembled in Malaysia. Two billion components leave our Malacca plant [image of plant comes on] each year. A product group manufactured here on a large scale are LEDs and detectors.
>
> [Images on screen] At this point, the sequence of images that appear consist of a quick montage of clips juxtaposing traditional images of the island such as a rice field, its lone sky-scrapper, a *kampung* house, a fishing village, wooden bridges extending out to sea, and hawkers *against* images of the plant showing assemblers at work, its machinery, clean rooms, and the technology; flipping swiftly from one to the other; the sound of local drums increasing in intensity and taking over from the narration during this time.

What is significant here is less the specific information conveyed than are the images and juxtapositions that appear on screen. While the actual narration gives credence to previous claims about the interconnectedness of production processes within the company the world over [*chips produced in Western country assembled in Malaysia*], the juxtaposition of images which follows silently constructs for the audience, a message singularly *modernist* in its constitution of Self and Other.

The kinds of images forwarded as representations *for* Self and Other are especially telling. We find, as an example, the epitome of modern efficiency and technological proficiency when it comes to constructions of the Self, albeit based on images derived from the plants in Malaysia: underlined here by images of assemblers garbed for high-tech laboratories, clean rooms, sophisticated machinery or technology and so on. By contrast, images representing Others forward and reinforce wide-ranging assumptions, indeed images, of them as underdeveloped entities. While clips of fishing villages, kampung houses, quaint little wooden bridges extending out to sea amidst a tropical sunset, hawkers frying noodles, and rice fields may, under certain conditions, be constituted in non-modernist or nonhierarchical ways, when juxtaposed specifically against the high-tech atmosphere and surround of the plants, these images imprint on the minds of the audience an Other who is far behind the super-glossy world of the electronics industry.

In the process, these juxtapositions forward an implicit message that the company, through its [re]-location of production activities in Malaysia, is very much *involved in* developing the nation—referenced here by images of the high-tech Self which came *not* just from anywhere, but from the company's plants *in* Malaysia. In this instance, what is brought silently to mind are tenets of neoliberal development discourses connecting the foreign direct investment (FDI) policies of multinationals to the development of newly industrializing nations like Malaysia.

What I have documented in preceding sections is the effective production of subjects as workers, as developing entities, and the specific constitution of relations connecting these subjects to the terrain of the modern organization, the economy and the nation. In particular, the assessment and regulation of the self that is endemic to the practice of quality management construct for these subjects the effectiveness, superiority and relevance of these precepts and practices to their own lives, be it as worker or member of developing nation. It is in this sense that self-assessments are "good for them." What needs noting at this juncture is how the discourse of quality simultaneously provides the discursive entry-point for acting on the local terrain, while also producing the very constructions of local officials and managers documented above. In this regard, the discourse, its concepts, practices and so on are constantly organizing these individuals' construction of problems (e.g., youthful and undisciplined subjects, lack of education and skills) and solutions (e.g., training, self-assessment and regulation).

THE KNOWLEDGE AND PRACTICES BEING PUT INTO PLACE

Central then to the interest in quality is the claim of such know how to build analytical and participatory capacities in people, groups and organizations. Constituted by a dual focus on quality assurance and reliability, the effective regulation of organizational activities, behavior and materials are seen as central to organizations' ability to be successful and competitive. In Infotech's case, quality assurance gates or checkpoints, sample testing, reliance on a battery of tests governing its products, and training workers to be proficient at meeting or undertaking these requirements are all examples of what is entailed.

However, it is not just training workers *for* participation that is important, but training them *in* the proper ways of participation. This need to regulate workplace behavior and activity was likewise noted by Ong (1987) in her discussion about the various modes of industrial discipline organizing women factory workers in Malaysia. In this instance, organizational requirements for conformance and tracking mean that there is an equal amount of attention paid to *how* workers participate. As I was to discover, there is a high degree of standardization to the methods and procedures of analysis used by quality control circles (QCCs) that is crucial for making possible this need for a regulated participation. Most come from quality control packages available for purchase by companies for training workers. In the case of the NPC and SIRIM, many of these tools and procedures taught by trainers were developed by the Society for QC Technique Development in association with the Union of Japanese Scientists and Engineers.

Concurrently, training packages for the civil service are part of a standardized package developed by the National Institute for Public Administration (INTAN). Relying on what is called the "cascade principle" (Sarji Abdul Hamid, 1994: vii), the dissemination of TQM precepts and training is facilitated by a "system of training franchise" (: vii) where:

> . . . standardized training manuals on the various [government, departmental] circulars [on quality improvement, quality control circles, project planning and preparation for example] . . . [are] prepared by the National Institute of Public Administration (INTAN) [for use] by other public sector training institutions and departments (: vii).

Through this cascade principle and training franchise, not only are training material and information standardized across the board for all civil servants, ensuring high "uniformity in instruction" (: 12); the deployment process is also more far-reaching in its actual dissemination capacity.

Knowledge and the Practical Constitution of Subjectivities

For both the public and private sectors, actual training typically involves some form of "theory and practical training." For Multitech, the continuous dissemination of information updating workers about quality activities is also deployed during its weekly TQM communication day. An almost half-day event attended by staff and workers alike [on a rotational basis in the case of shift workers], highlights for the event include updates from senior executives about TQM-related results or goals such as business situation updates; detailed project presentations by different quality control teams, including cross-functional, systems improvement, and self-managing teams; and finally, introductions to new quality management or benchmarking information like Kaizen from the TQM Training and Development unit. Infotech activities and practices are equally saturated by this focus on quality. Monthly quality excellence reviews and communications involving all administrative and production units are organized specifically to obtain and share operational and quality updates. This is paralleled by weekly staff meetings between individual units and the quality control department.

In general, information about a variety of standardized tools and techniques such as the 7 new QC tools, charts (milestone, bar, radar), and diagrams (pareto, fishbone) are the practical arsenal taught to workers. As part of the training process, participants are divided into groups and asked to discuss a problem using these tools and techniques, the coordinator or trainer going around to help members as they do so. The intent here as told to me by one of the NPC trainers, is to "get them to act like a QCC group" right from the start. Facilitator training packages usually take about 3 to 4 days, although follow up and support activities may last as long as 3 to 6 months.

From the standpoint of knowledge and power, this active regulation, dissemination and application of quality know how taught to workers is at heart a story about the practical constitution of reality and subjectivities (Bourdieu, 1977). Effected and mediated by a variety of conceptual, textual and documentary forms, the power of such knowledge to regulate and constitute resides in their effective ability to constantly shape and organize local workers' apprehension and sense of self, of reality and of relations between this self and knowledge, self and others as well as self and the work/workplace. Specifically, the coterie of schemata, categories, causal relations and rationality that are encoded in the aforementioned tools and procedures are central procedures by which workers' apprehension of self and so on are institutionally and textually inscribed. In the words of Dorothy Smith (1987), what is being organized here is the active and planned dissemination of those "ordering and structuring procedures" deemed necessary to sustain workplaces and institutions of one kind or other.

An overview of the aims and objects of techniques taught to workers pinpoint attention to this implicit and explicit regulation of sight, self and relations. As an example, according to a training flyer produced by the NPC and SIRIM, techniques such as the "fish-bone" diagram and the "7 QC Tools" [a composite of diagramming procedures], are intended to provide workers with systematic ways to "clarif[y] causal relationships," "organize problem-solving data in[to] appropriate groups," "derive appropriate means to accomplish objectives," "arrange numerical data," "chart process decision programs" and "establish daily plans for time-related events" (NPC and SIRIM, 1995). In this instance, the object is to develop workers' ability to "clarify," "organize," "derive," "arrange," "chart" and "establish" information into desirable and specific forms of knowledge (i.e., objectified, functional, hierarchical) that are subsequently amenable for worker, managerial and/or organizational action and intervention—articulated here in terms of "problem-solving," "decision program[ming]" and "daily plan[ning]" needs.

Meanwhile, in the case of the widely used fish-bone diagram [also known as the "cause-n-effect" tool], what is given to workers is a specific procedure for identifying problems and their causes that logically and systematically reduce reality into functional and distinctive conceptual categories. Intended so, the procedure *itself* illustrates the already predefined constitution of reality that are given to workers. From the outset, workers are given four potential categories/causes—man, machine, method and material—that serve as guides for them as they seek to identify and assess whatever problem it is that they are encountering in the conduct of their work.

I will use a specific example of the cause-n-effect process used by a quality control team from Infotech to illustrate this systematic constitution of worker activity and logic referred to in the above. In this case, the group was concerned with trying to identify the causes leading to "scratches on LCC units"—an outcome often leading to the costly rejection of these defective parts [see Figure 5.1]. Calibrated and categorized *in terms of* the four predefined categories given by the tool, potential reasons for each of these categories are systematically identified and verified by team members as to their status as causes. We find at this juncture, attributions such as "discipline" and "no follow-up" as potential man-related causes; "dirty combo" and "low quality combo" as potential material-related causes and so on. In this specific example, verification [a process mediated by observation and inspection via the activity of counting], leads to the conclusion that "scratches from the test handler"—listed under the category of machine—is the probable cause.

This preliminary *deduction* leads to the second level/stage of analysis [see Figure 5.1] focusing more specifically on "scratches at [the] test handler" that is now taken as *the starting point of analysis.* At this point, the cycle of identification and verification of probable causes in accordance to the four categories start all over again. In this example, this progressive deduction(s) led eventually to the conclusion that scratches on the LCC units are caused by the "overly strong force of the singulator spring" attached to the machine [not shown].

On the basis of such analysis, suggestions are given on how problem[s] may be resolved. Trial runs and documentation of these trials are often undertaken at this point. It is a process of evaluation involving yet more cycles of identification, documentation, analysis and so on. Often included here are considerations about the myriad of pros and cons attached to suggestions for fixing the problem (e.g., cost, jamming potential, handling procedure, training), along with documentation of results and cost-savings where applicable. For this particular group, estimated cost savings attached to the project was calculated to be US$525,025.00 per year.

(1) Initial Identification and Analysis

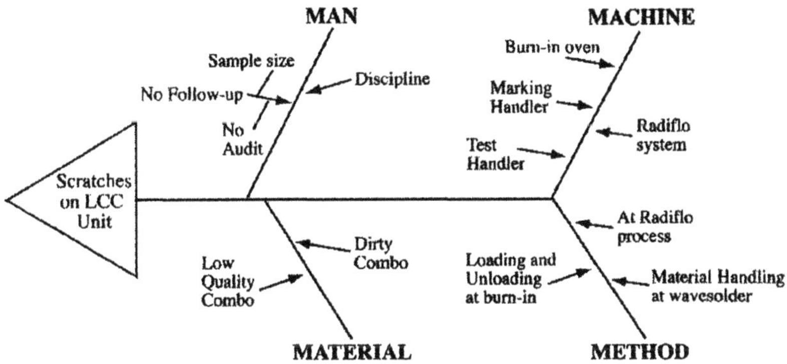

(2) First Level Analysis

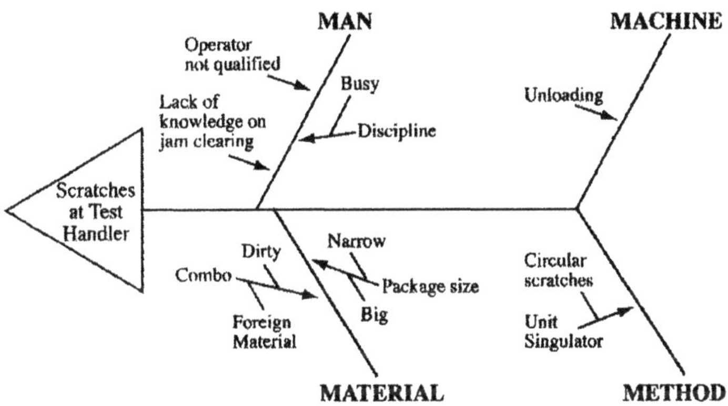

From the Regional Quality Control Convention organized by the National Productivity Center, July 8–9 and July 11–12, 1995, Kuala Lumpur, Malaysia.

Figure 5.1 Recreation of a Quality Control Circle Analysis Using the Fish-Bone Technique

Thus, embedded in the fishbone analysis as presented here is the provision not only of very particular categories *for* analysis, but a systematic process *of* analysis as well. Possible causes identified and organized via the textually mediated categories such as man, machine and so on become problems unto themselves that are subject to further scrutiny. It is a sequential process and logic whose aims are the systematic identification, investigation and documentation of a seemingly never-ending series of problems-causes-and-solutions. As a result, one may find, as in the example used here, multiple iterations and use of the diagram throughout one single presentation.

Informing and aiding workers in their effective ability to identify and isolate factors/causes are results [and knowledge] gleaned and derived from their use of the aforementioned "7 QC Tools." Specifically, these tools provide workers with supportive techniques they can use in order to organize information gathered or problems encountered into factors that can then be converted into the various cause-and-effect relationships they need to conduct the fishbone analysis. That is, relying on the objectivist and reductive form of logic embedded in these tools to organize and make sense of information they are gathering to represent the four fishbone factors—a procedure reminiscent of Adelle Mueller's (1987b) notion about the "power of naming"—the systematic conversion of factors into identifiable causes are mediated by a series of predefined causalities [represented by where arrows are pointed] that functionally links them to whatever outcome or problem workers are focusing on. Thus, what is seen or represented by the fishbone diagrams are more often than not, already mediated by the categories, deductions, causal attributions and so on encoded within *these* tools as ways of knowing.

What is important to note is how these practices of clarifying, organizing and charting are able to organize workers' reality *precisely* because of the already-defined nature of the tools and schemata taught to them: reductive, functional and linear. As such, when workers identify or assess problems in their everyday world, they do so mediated by the categories and rationalities already encoded in these conceptual and documentary forms. As an example, when applying the fishbone diagram, not only is attention systematically and consistently drawn to the four predefined categories, each has already been constituted by a form of logic and constitution of reality which allows workers to systematically link *and reduce* phenomenon observed to these categories, and iterate from one level to another.

Given the above, it is not surprising to find that there is often a great deal of consistency in how categories are used or constituted by workers. In multiple QCC presentations that I observed, "man" is often defined in terms of people's rates of participation, training (e.g., skill level), or attentiveness (i.e., whether workers were rushing or careless at work). Concurrently, "method" is often represented by smoothness of the workflow or coordination procedures, whether calibrated in terms of interruptions, cycle times or confusion over the clarity of processes; "machine" by turnaround time or breakdowns; and "material" by acceptance rates, defects or availability. The last two in particular are often defined in highly quantified terms.

What we find replicated in workers' constitutions of categories and factors are those very forms of knowing informing officials' and managers' constructions of subjects and reality: that is, filtered by a conception that is economized and objectivist in nature, concerned with productivity, efficiency, worker discipline and so on. Moreover, not only are these forms traceable to the discursive heritage framing these tools and techniques—technocratic, rational, economized—they are also mediated by the organizational interests these tools and procedures are there to serve. This ability and capacity of workers' to [re]-produce a form of knowing and doing that is functional, rational, efficient and linked to the interests of the organization is what makes these workers and their participation relevant to both workplace and economy.

However, by doing so, workers and group members are constantly involved in socially organizing their own reality, their work activity, their relation to this activity, to their co-workers and so on. What is affected in the process of people acquiring, learning and/or applying these tools and schemata, logic and rationality, is nothing less than a *functional constitution of reality.* That is, an instance of what enculturation into the vagaries of becoming knowledgeable, independent and self-regulating subjects of the modern economy is about.

* * *

Malaysia in the midst of change is a terrain at once strange and familiar. While reminders abound everywhere that I am in a place I had once called home, the overriding impression throughout my first few months there seemed to be one steeped in the vagaries and vicissitudes of modern life. I suppose what I think of now is less the more obvious indices of change that first grabbed my attention than I do the impact and meaning of such changes to life as it is lived on an everyday basis in the nation: the traffic congestion and rush hour madness; teeming skyscrapers; McDonald's; and oh yes, Calvins too—Klein that is. Distinctions and symbols that are associated, rightly or wrongly, more with home-away-from-Malaysia than Malaysia itself. It did not take me long to realize the magnitude of change that is the nation now. In truth, it took me less than half an hour into my first meeting with one of the managers from Multitech—a scant twelve hours after my arrival in Penang—to realize what change entails.

"What do you mean? There are no snakes in the snake temple? How can there be a snake temple without snakes?" I am speaking about snakes and temples and temple-less snakes or, to be more precise, snakeless temples, in a multibillion-dollar conglomerate. Shades of a twilight-zone-like reality or irony, whichever you prefer, whose naturalness is embedded precisely by the existence of said conglomerate itself. But let

me not get ahead of myself too much. "There are no snakes." So came what seemed like a very definitive answer. A shake of the head and a chuckle seemed only to add insult to a very personal injury. "But—it's the snake temple! I saw them! There used to be tons of them, they were everywhere. [Pause] What happened? Where did they go? Are you sure?" ". . . Well, there may be a few of them here and there but . . ." But?

Or as his manager grinningly explained to me some hours later as we all sat down for a late lunch: "Oh, they ship them in nowadays, Vanessa. They put them in these crates and ship them over from Thailand or somewhere like that and they release them here. The tourists come, they see the few snakes, and they ooh and aah." It took a while before I realized he was pulling my leg. Chuckling reluctantly as I shook my head, I did not know whether to feel like a "villager" or a tourist or both. I decided on both.

And the snakes? Most are gone, as both men had noted. Maybe dead or maybe having beaten a hasty retreat to safer haven as the woods that used to surround the temple had long been cleared to make way for development. Or to be more specific still, to make way for the development of the Bayan Lepas Free Trade Zone.

Today, standing right next to the temple, you will find a gleaming steel and glass structure with the name of some multinational on it. Funny, I had been so excited on realizing that my research site in Penang was so close to the temple, I never even saw the temple, much less the plant next to it, when I passed it by on the way to my site. It is surrounded now by concrete and glass. With food stalls catering to nearby factory workers located slightly to the front and side of it. And running like a gauntlet to its entrance, yet more stalls hawking cheap trinkets to eager tourists pouring out of tour buses.

I never did bother to go to the temple, or found out where the snakes had gone. Although I have to admit, I did find myself standing outside the temple for a minute or two one day, memories of my last visit there, some twenty years ago, flooding through me, flanked on both sides by the touristy stalls, undecided whether to proceed or not. Finally, I turned away. Somehow, it did not seem to matter. Or maybe it was just fear which kept me away; of what I would see—or not see. I scarcely know which. Funnier still, I did not even think of the temple, much less saying goodbye to it, when I finally did leave Penang. I had forgotten.

* * *

COORDINATING INSTITUTIONS AND REGULATORY FORMS BEING PUT INTO PLACE

Whereas multinationals usually have their own in-house quality and QCC training programs and facilitators, institutions like the National Productivity

Corporation (NPC) and the Standards and Industrial Research Institute of Malaysia (SIRIM) are central nodes for disseminating quality control training to local small and mid-size industries (SMIs). A designated site for monitoring and conducting research on productivity and quality in the nation, the NPC is charged with "upgrading local expertise" and promoting productivity and quality activities for both the public and private sectors although its priority remains focused on local SMIs.

Initially set up in 1962 as a joint project between the United Nations Special Fund and the Federal Government of Malaysia, the National Productivity Center [as it was known then] was intended as a research and information gathering site on matters pertaining to productivity in the nation (NPC, 1994, Box 1: i). With the International Labor Organization (ILO) as its executing agency, its aim was to introduce the concept of productivity to the nation, as well as to aid in setting up an infrastructure—institutional, administrative, conceptual—supporting the use of the concept as a barometer for development planning and measurement.[3]

* * *

But temple grounds making way for factories and whatnot is not the only instance of how the country is changing as it strives to achieve and continue its eight-percent-a-year growth rate. Everywhere that one looks, there is something under construction—or destruction. From physical infrastructure to the environment to social life to community identity to corner newsstand, the cycle of creation and destruction is there to be seen or ignored.

There were times when it seems like scarcely a week would go by where reports of yet another catastrophe dealing with the dumping of toxic waste, fatalities or accidents at construction sites, or dislocations of communities and lifestyles were not reported by the local news media. Often paralleling such occurrences were the equally disturbing news of abuses the million or so foreign workers were subjected to: from poor working conditions to shabby treatment by employment agencies to a myriad of social problems both individual and national.

I think now as I did then, as indices of economic growth, what do such changes signify for both polity and nation alike?

* * *

With the passage of the National Productivity Council Act A801 1991, the concept and practice of quality management became an indelible component of the newly incorporated institution's overriding concern with productivity (NPC, 1992: 40). Constituted simultaneously by the discourse of global competition and a growing concern with labor shortage[s] at the local level, this active association of quality *to* productivity was an attempt by state planners to confront these twin concerns via attention to certain "productivity-linked approaches to competitiveness":

> The traditional approach of relying on low-cost labour, abundance of raw materials and on low technology or labour-intensive operations will neither be conducive nor competitive in a situation of labour shortage and rising labour cost. A better strategy would be to adopt a *two-prong approach* to enhance productivity *and* quality [which includes] . . . maintaining a suitable balance between investment in productive physical capital to increase production capacity and continuous investment in human resource development to enhance the quality, technical and managerial skills and capabilities of the workforce (NPC, 1994: iv).

With this reconstitution to its aims and processes, the NPC broadened its basic productivity research and monitoring activities to include a more active role with respect to actual training: in effect, becoming a central organizing, training and consultancy agency in matters of productivity, quality planning and development for the local private sector and the state.

Linked thus to the HRD objectives of the state, "P and Q enhancement programs" within the NPC include a variety of courses both pedagogical and practical for consumption by local administrators, managers, supervisors, entrepreneurs and workers (NPC, 1995). Included in the former are courses like QCC Appreciation Seminar for Managers, and Managing Resource Productivity in Manufacturing. For the latter, courses include QCC for Facilitators, QC Tools and Techniques, and the 7 New Management Tools for TQM. In terms of actual training, numerous "in-plant sessions tailored to the needs of specific organizations and companies" are a key form of deployment. Interested companies approach the NPC, who trains QCC facilitators for these parties. Once ready, they go and start their own quality circles. At the time of my visit, there were approximately 2,000 quality control circles registered to the NPC.

For the public sector, institutions like the Malaysian Administrative Modernization and Planning Unit (MAMPU) and the National Institute of Public Administration (INTAN) are the agencies charged with deploying such knowledge and skills to civil servants across a variety of departments and institutions, including municipal councils, hospitals, universities, employee organizations, state departments and government ministries (Sarji Abdul Hamid, 1994). Working closely together, they are responsible for identifying administrative, planning and quality-training requirements deemed necessary for modernizing the civil service. Often constructed under the neoliberal auspices of deregulation and administrative efficiency, the focus on TQM became an indelible component of civil service training with the introduction of the Client's Charter in June 1993, although numerous Development Administrative Circulars pertaining to the management of quality were issued as early as 1991 (Sarji Abdul Hamid, 1994).

Coming under the jurisdiction of the Prime Minister's Department, MAMPU officials often go around the country organizing and giving talks on quality improvement and quality control circles to various federal, state and municipal departments and units. By contrast, INTAN is the institution charged with deploying the actual training activity. Noted earlier, it is the agency responsible for making-real the government's concept of a training franchise, developing a standardized training package for use by all public agencies (Sarji Abdul Hamid, 1994). In 1993 alone, a total of 10,500 civil servants were trained by INTAN in the various approaches, including quality management and QCC training.

Institutional Mechanisms for Regulating Participation and Change

A variety of institutional and administrative mechanisms are likewise in place to promulgate and encourage participation in quality management and quality control activities, including incentives, grants and requirements for trade. As an example, local SMIs are encouraged to partake in these training opportunities at institutions like the NPC through the provision of rebates and double deduction incentives by the government. Moreover, under the Human Resources Development Fund (HRDF), employers registered and/or incorporated in Malaysia who register with the Human Resources Development Council (HRDC) and contribute to the human resource development levy for a period of six months are eligible to apply for training grants and financial assistance to defray part of the costs of training their employees (NPC, 1995: iv). Under this program, eligible companies can participate in a variety of development schemes, including schemes for training, retraining and upgrading the skills and knowledge of employees and for participating in any publicly offered training program.

Administrative and financial inducements encouraging SMIs to undertake a variety of industrial research and product development activities are also in place. Initiated "to complement the Government's efforts to assist small and medium scale industries to modernize [their] operations and . . . become self-reliant" (SIRIM, 1995: 27), the Industrial Technology Assistant Fund (ITAF) is an example of incentive programs set up to encourage and transform local SMIs into modern and efficient entities able and capable of competing in the international marketplace.

Under the ITAF's "Quality Improvement Program," all small-scale industry wishing to participate in the Ministry of International Trade and Industry's (MITI) vendor program are required to pass its quality improvement criteria. This process often involves introduction to, training in and adoption of quality management standards and practices by participating companies. As noted by an officer from MITI, the program had been formulated to ensure that all participants "meet a certain level of standard." Between 1991 and 1994, 183 such ITAF schemes were approved by SIRIM (SIRM, 1994: 57).

* * *

Often, economic growth itself seemed to be an enormous double-edged sword, even to those proponents whose voice can often be heard advocating the continuous reliance on such policies. The shortage of labor engendered by the growth in the nation is a particular case in point. Often leading to calls to either increase the availability or accessibility of foreign workers, reports such as those noted previously lead to a sometime vacillating policy by state agencies. Torn, it seems, between the need to meet such calls and so sustain growth on the one hand, and by the equally pressing need to confront how continued abuses and/or increasing social resentment against the presence of these foreign workers can be balanced or dealt with on the other

Concomitantly, the increasing sense of well being engendered by the nation's prosperity seemed to silence any overly harsh critique of both changes and policies alike. At least in so far as advocates of its development policies are concerned, whether local or global. For the general polity, life seems to circulate around an increasingly busy and seemingly never ending cycle of work-and-expenditure.

Indeed, my overriding impression of "Malaysian society" on first encounter was an awe-like wonder at how consumerist people had become. "There is rampant materialism gone amok here in this country": so reads one of my earliest reflective entries. From cars to houses to clothes to the ubiquitous hand-phone [irritating status symbol which it seems to be here]: everything is up for grabs, nothing is left out, nothing overlooked. And brand consciousness? It was the "name" of the game.

Levi's, Chanel, Armani, Esprit, BMWs, Alfa Romeos—all were taken as barometers for, and symbols of status and character. Identities, it would seem, are now to be connected to one's purchasing power, and status derived from one's ability to locate the self as a modern market citizen. And it is only now as I reflect back on the exigencies of the nation in transition, that I think to myself: "how fitting, the consumer to go along with the economically relevant worker." At the time, I had been too much in the midst of "living" the reality of a Malaysia constantly going shopping to really make the connection.

As the agency vested with the regulatory responsibility of establishing and monitoring quality and industrial standards, SIRIM is centrally involved in efforts to regulate the activities of local institutions. A statutory body set up to "enhanc[e] [the nation's] international competitiveness through partnership in industrial technology and quality" (SIRIM, 1995: 10; see also SIRIM, 1994),[4] the Institute's main objective is to support some of the industrial sectors targeted for action under the nation's Industrial Master Plan (IMP) via research activities and the setting or regulation of industrial standards. In its regulatory and monitoring capacity, information and standards established by SIRIM often become the barometers and criteria by which programs, loans, grants and so on are approved within other agencies.

Moreover, as the certifying agency for the International Standards Organization (ISO) and other certification bodies,[5] SIRIM is centrally involved in organizing and mediating between global industrial requirements and local compliance. In this instance, it is a process of regulation that often entails fine-tuned monitoring or analysis of local companies' activities and practices in areas like administration, materials handling and production, training and skills development.

From the standpoint of knowledge and power, these requirements for certification are a powerful administrative procedure for regulating the adoption and implementation of industrial and QC know how constituted and coming from *elsewhere*. Or, as formulated by the aforementioned official from SIRIM in his explanation of why local industry is so interested in all matters quality: "the reason TQM is so popular is because of compliance with ISO 9000." In this instance, it is a reminder that such certification requirements are themselves framed by, and already connected to a political economy that compels local participants in the economy to comply in order to remain viable or to even survive. A specific example of such an implicit disciplining mechanism is quality standards imposed by importing countries [or corporations] as a requirement *for* trade with their local counterparts.

Thus, institutionalized as part of the administrative and regulatory activities of institutions like the NPC, MITI and SIRIM are a series of financial, administrative and regulatory procedures such as rebates, grants, standards of approval, requirements for participation and so on *whose* existence are wrought to both compel and instill within local SMIs, very particular types of industrial or market activities, and practices of quality control and teamwork. Directly connected to the state's aim of producing and effecting a viable local industrial and entrepreneurial sector, these programs are effective institutional forms mediating and organizing the activities of local companies. Taken together, the activities of these institutions configure themselves into an interlocking web of practices connected by close administrative and procedural ties.

From the standpoint of institutional ethnography, what needs noting is how such requirements and interconnections—institutional, administrative—come already socially organized and framed by very particular intentions and beliefs produced by discourses currently circulating in the nation. In this way, the discourse of quality and its articulation in the nation's development objectives is, and has been, a major organizer of institutional activity and life during recent times.

While less focused on financial inducements compared to the private sector, a variety of awards and quality inspections are key procedures of control and regulation governing the activities of civil servants and the public sector (Sarji Abdul Hamid, 1994: 11). Under the Client's Charter introduced in 1993, all government departments are not only required to have a written assurance as to the quality features and standards of the product or service they offer to their respective customers, they are also required to "ascertain that these features and standards generally match with the clients' expectations" (: 11). According to the head of the civil service:

> [u]nder the *Total Quality Management Programme* all public sector agencies are required to formulate their respective missions through the process of strategic quality planning. They are to be *market-driven,* and to institutionalize a distinct *customer-orientation* in the delivery of services (Sarji Abdul Hamid, 1994: iv—v).

Where failures to comply with such expectations occur, a "service recovery mechanism" is available within the respective agencies for undertaking "remedial action" (: 11). As a result of such changes, civil servants are not only required to be more efficient and productive in their delivery of services, their very activity and work are now explicitly linked to the market and economy; their everyday activity and work procedures subject to the scrutiny of a phalanx of individuals, both institutional and non-institutional (e.g., colleagues and supervisors, quality inspectors, the public).

Much like the case of women factory workers whose everyday behavior and conduct was systematically and progressively opened-up for surveillance as a result of their shift into the terrain of waged labor (Ong, 1987), what we see occurring here is the increasing presence of new forms of surveillance and discipline over workers and the workplace. Civil servants and their private sector counterparts (e.g., young Malay operators, local SMIs) were largely overlooked in terms of their day-to-day activities regarding the conduct of work, organization of space, and use of time until this reconstitution of their subject-location as valuable development resource. Today, the application and utilization of QC precepts and practices to the terrain of modern workplaces, both public and private, have opened-up to scrutiny the everyday behavior and motion of these individuals and companies in their location as *workers* or *terrains of production*. This is a "scopic possibility" and a visibility directly linked to the neoliberal, market-oriented gaze of modern discourses circulating in the nation.

* * *

> *Assessments of good and bad, desirable and undesirable are thus increasingly associated with material wealth or the show of it, anyway. As noted by an uncle in commenting about such changes occurring in the nation, "even the poor says, "no matter how wealth is attained, there is no shame worse than poverty.'"*
>
> *A phenomenon often associated with the notion of "wanting face" in Hong Kong, and "kiasu" in Singapore [a Hokkienese term literally meaning "fear of losing"], this focus on consumption and material well being seems to be more preoccupied with fear of losing than it is about maintaining a good show—a competitive element somehow having crept into it as a signifier for changing notions of self and relations. Cellphones used in the most inopportune of places, with loud conversations heralding back to the person's favorable financial position becomes unto itself a statement of being, of "having arrived." I wondered to myself if this is the destination that development's "kam-ap" [take-off] had been bounded for all along. What would grandfather have made of all this?*

* * *

Central Nodes of Mediation

Institutions like the NPC and SIRIM are thus important sites or conduits for both *easing* knowledge *into* and redeploying this knowledge *out* to the rest of the nation (e.g., industrial standards, QCC practices, quality management). In this mediatory capacity, they also serve as important sites for convening interactions and exchanges between representatives from a variety of sectors (e.g., local SMIs, foreign MNCs, government, academe), and professions, including management, engineering, science and technology. For example, as the administrative body for the Prime Minister's Quality Award, the NPC is a funnel for bringing together representatives from government and industry as evaluators for its examinations and judging panels, and/or members for its board of directors. Winners of quality awards [mostly foreign MNCs] are required, under the terms of the award, to take on the role of quality counselor and speaker in a variety of public service-engagements aimed at informing and sharing with other aspiring organizations [notably, local SMIs], the specific features and processes involved in transforming *their* organizations into potential winners of like endeavors.

I bring attention here to the presence of these mediatory institutions because they constitute a crucial part of an interlocking web of private and public, local and global relations and discourses, whose practices and administrative procedures are centrally involved in organizing and reorganizing the local terrain. It needs noting at this juncture that the connections [epistemic, administrative, economic] effected between industry and government in their joint effort to constitute and deploy quality know how is but one example of multiple links forged between them. Examples of other connections sought after or formulated by the state include the active cultivation of multinational participation in a variety of other educational and training endeavors.[6] Included here are skills development centers, sponsorship of entrepreneurial training programs in schools, industrial or organizational internships, and other contributions in more social and cultural events. In the case of these skills development centers, the land and structures are provided by either state or federal agencies, and the technology, instructors, strategic planning input and so on, by MNCs and academia [on certain occasions].

Like the NPC, SIRIM is a conduit between the government and industry, and a site where information from elsewhere are systematically gathered, assessed and dispersed to local clients, and to a lesser but growing extent, to international clients such as various South Asian, African and Southeast Asian nations wishing to set up SIRIM-like entities or who are in need of training facilities and support. Concurrently, the organization of talks, research forums, seminars, workshops and so on by both agencies provide a forum where foreign expertise and know how are introduced into the nation.[7]

Thus, the NPC and SIRIM function in a capacity akin to Smith's (1987: 160) notion of institutions as "nodes" or "knots" where fine webs of knowledge and power are constituted and affected, producing a[n] [interlocking] web of knowledge, of administration and so on aimed at "coordinating multiple strands of action into a functional complex." Moreover, it is not just a simple focus on administrative coordination or institutional collaboration that is significant here. Rather, the focus is on *how* such coordination and collaboration are textually and administratively mediated by systematic interconnection[s] between forms of knowledge and techniques of power: for example, planning and administration, criteria establishment and evaluation, standards and certification. In this instance, interconnections themselves are forms of knowing and doing already socially organized and belonging to very particular institutions of rule.

As noted by Escobar (1994b) and others (e.g., Ferguson, 1990; Mueller, 1987a; Sachs, 1992) in their deconstruction of the development apparatus, understanding the link between various development efforts and the construction of new subjects entails moving *beyond* the simple assessment of such policies and programs *in their own terms*. What is called for instead is a focus on these links as the "coordinated work of people" (Mueller, 1987a; Smith, 1987)—state planners, development consultants, managers—whose activities and efforts are aimed at specific subjects (e.g., young Malay workers, civil servants) who, in their location as objects *for* intervention, are as much a *product* of the discourses and professional training of these ruling individuals as they are an index of some "naturally-preexisting" social condition or biological predisposition (e.g., Malay backwardness, underdevelopment).

* * *

With the nation's economy seemingly growing without end, and international recognition becoming a facet of everyday life, rampant and boundless optimism makes possible the illusion that one can have it all, even as people are starting to pay a price—longer work hours, organizational claims on identity and self, stress—for all this "good living" [as a local journal aptly satirized]. As in the soap operas from Hong Kong followed so assiduously by big segments of the Chinese community in particular, people are inundated from all sides with advertising, movies, and seminars about what this life is, and how happiness is or can be bought. So much so that statements from the young like, "he or she's very branded, you know" becomes a common way for communicating character, position, status and so on amongst themselves and to others.

The statement above was one actually made to me by a little cousin of mine in her attempt to convey to me the eminent attractiveness of a friend she particularly admired, unknowingly plunging me through a roller-coaster of feelings both gruesome and surreal, even as outrage and horror did emerge pretty quickly. "Do you know what "branding" means? It's about the "giving-up" of self, a statement of ownership that claims one for its own. Cows are branded, not people. . . ." Silent apology to cows notwithstanding, this little delivery is followed quickly by an overriding sense of contrition as I realized just how "branded" I myself am, hardly the model of pious deprivation and sobriety, to say the least.

Why did I never see this particular genealogy of branding before in my own life and life in the West? In this context, I think too of how people were literally branded in the name of economics and purity, subjects of hatreds unimaginable but all too real—double-edged swords and paradoxes indeed. How do I as a researcher and social critic [such as I am], not to mention cousin, highlight this and maintain the integrity of my criticism without losing sight of my own implication in the process?

* * *

Connected to this [re]-constitutional possibility is attention to who is speaking. Noted in the introduction, the world I traversed during my time in the field was a world peopled by technocratic experts of one kind or another. Educated mostly outside the nation—in the United States, the United Kingdom, Australia, New Zealand and to a lesser extent, Japan and Singapore—and often the owner of more than one professional degree, these individuals I interacted with spanned across a whole range of modern occupations in the social and applied or hard sciences. As an example, most of the managers from the two multinationals I was researching have degrees in engineering and business administration. Likewise, administrators and officials from the statutory agencies and government ministries have degrees in a variety of social or management sciences, including human resource management, entrepreneurial studies, public administration and economics, although some have degrees in the sciences as well. In the case of a few senior officials in the public or semi-public sectors [not including universities], individuals have doctorates in economics, public administration and management.

As such, when these individuals assess and [re]-present their specific analysis of needs, events, solutions and so on to me, to their colleagues and to yet others elsewhere, the constructions they forward are *already mediated* by the technocratic, objectivist, scientific and modernist lenses engendered by their professional training. It is in this sense that attention to "who speaks" matters since this speaking extends beyond the boundary of speech to include actual assessments, practices and actions taken by these individuals in *their* conduct of everyday work, whether in identifying and planning training needs, or in formulating institutional and administrative programs.

Nowhere is this constitution as telling as it is within the academic community and institutions of higher learning in the nation. Much like their civil service counterparts, public universities had been subjected to this quality and market mediated gaze of state planners and administrators throughout my time in Malaysia. According to one of the leaders of a local university, in his response to my inquiry about this interest in deregulating and corporatizing universities in the nation, "basically, we want to make ourselves more relevant to the needs of the nation and of the market."

This conscious linking of the self to the market changes equally the very basis and relation that *organizes* the activities of these institutions and those individuals residing within them. Of particular significance here is the value that is now conferred, under the market-mediated discourse of quality and efficiency, on a form of knowledge that is *practical* and *linked* to the exigencies of industry and economy. Take as an example the following constitution forwarded by yet another senior official from a university:

> . . . once corporatized, [administration] may want to look at the kind of research one does—more hands-on, more practical, market-oriented, not the ivory tower type . . . there'll be more stress on external linkages, professional connections, collaborative research . . . local linkages with industry will also be important . . . we want faculty to generate consultation business . . . also to help reorient ourselves . . . after corporatization, assessments will be conducted on uses of our products. We need to find out what's lacking, what's not . . .

What is important now is knowledge that is practical and market-oriented, not pertaining to the "ivory tower." It is the focus on *use* that is significant. Academe and academics themselves must now learn to be better producers of products in the service of the market and nation, the presumption being that what is good for the market [or industry] is *also* good for the nation. It is, to paraphrase Mahathir Mohamad's (1993b) words, education for the shop floor and the factory that is important. The assumption informing these constitutions: the institution and activities of members residing in it should have a purpose and a functional aim. *Knowledge,* in other words, will now acquire its meaning in relation to the market.

In sum, by focusing on current interests in quality management as a discourse, and by attending to the systematic constructions of problems, needs and subjects as the creation of these [interconnected] organizational, professional and development discourses, it is possible to understand development activities not just as some "solution" to Underdevelopment, but as instances where subjects of development are themselves produced and organized.

* * *

> *Juxtaposed against this preoccupation with all things material and economic is the equally material and economic concern over disparities between rising costs of living and lagging increases in wage rates—especially so among the low and semi-skilled workers in the manufacturing industries or lower echelons of the civil service. Lack of a minimum wage policy, and the continued [or growing] reliance on cheap foreign labor acts, as some local journalists contend, as suppressors of these wages—a situation that has contributed partially to the increasing social tensions between locals and foreign workers.*

> *Then again, there is the rising incidence of [social] problems like "bohsia." Often of grave concern to parents, teachers and government alike, bohsia involves the offer of companionship or sexual service by young girls in return for gifts of material things, typically something trendy, branded and so forth. Connected to the focus on material ownership as a barometer of status, parents from a variety of classes and communities find themselves caught between a way of life they either uphold or cannot afford and a notion of being that is singularly materialist in nature.*
>
> *Concurrently, struggles with rising medical and educational costs, difficulties attached to locating good childcare and so on also serve as a reminder that there are issues and consequences attached to the nation's prosperity, that development extends beyond the realm of the economic, even as these issues and consequences articulate themselves mainly in an economic guise.*
>
> *What indeed, has been effected under the auspices of a market-driven model to development? What has been put into place?*

* * *

SUMMARY

The introduction of market-mediated discourses of development and the subsequent focus on quality know how, once predominantly to be found in the private sector, have been actively appropriated by state agencies to cover *other* arenas of social life in Malaysia. What is put into place and deployed is an effective apparatus geared at transforming both the local polity and institutions. There is now a greater focus on efficiency and a conscious linking of workplace activities and institutional aims to outcomes of a practical and market-oriented nature. In the process, the very role of institutions in society, and the subjectivities of members within these institutions are being constituted and re-constituted.

This implicit and explicit focus on a practical and functional knowledge, relevant to the aims of the market and the economy is a theme underwriting changes in relations taking place in the nation. Indeed, such a constitution is the basis for change itself. Accordingly, much time and attention is spent by individuals charged with administering or deploying such changes as are occurring in the nation, constructing precisely the relevance of such knowledge. This focus on a functional knowledge—and implied revaluation of knowledge—highlights that issues at stake are not just epistemic, but *economic* in nature.

This reminder about the connection of knowledge to institutional and economic aims came across loud and clear during a conversation with one of the senior staff managers at Infotech. Distilled to its most basic, the message conveyed is one that stipulates that all knowledge, unless practical, technical or applied in nature, is of little relevance to the company and its objectives:

> . . . we like this, you see [a book they had been using to develop performance measures for participation] . . . you can actually *do something with it* what, [pause] not like some of the other writings . . . we had this guy come in a while ago . . . we told him, "Your theories sound fine, but what can we do with it? [pauses, shaking his head] it's *not* practical" [emphasis] . . . he was quite upset with us*lah* . . . we told him we couldn't see any practical use from it . . . there has to be practical implications isn't it?

As exemplified by this statement, in the constructions forwarded by administrators and managers alike is a specific valorization of what is deemed to be important. Observations regarding desirability and non-desirability anchoring managers' and officials' earlier assessments of workers extend here to the terrain of knowledge itself.

* * *

> *The effect of such change is in some ways best illustrated by the following encounter, where current interests in all things "quality" and "market-related" were brought to my attention good and proper. During a visit with an academic at one of the local universities, I was invited to join this individual in a function at the school. While there, I had the occasion of being introduced to a former colleague of this individual's, newly transferred to a management school. A few minutes into our introduction, I was told that he is currently working on a project on "TQM and children."*
>
> *At my blank countenance and whispered inquiry to the side [what is this about TQM and children?], I received the following exasperated response [also whispered]: " . . . people are going into schools to teach the children about TQM . . . how to manage their time, organize their work. . . ." Indeed, I was so flummoxed by what this person was telling me, I missed some of what was said, as the import and implications of what was being conveyed to me sunk-in—much like a bad case of indigestion.*
>
> *The unflappable researcher proved at this point, to not be so unflappable. "Oh no, I'm really depressed now . . . b-b-but . . . wh-what . . . how can they do that?!" What seemed like the longest pause followed, then came the response: "I probably shouldn't tell you, you'll be even more horrified . . ." "I know, I didn't know what to say . . ." Chuckling softly still at my reaction and lame response, I was told that most of these efforts are typically formulated or implemented under the guise of "Bumiputera Development."*

Individuals involved thus usually start by tabling their proposals to local schools, go on to introduce and/or run a course or two before writing yet more formal proposals. As my colleague goes on to note, such efforts are often difficult for teachers and schools to refuse, and after a while, the parents are crazy about it. After a brief pause and more reaction on my part, I was informed that this is one way through which locals start establishing themselves as experts and consultants, making a mint along the way.

* * *

Extrapolated to the context of a Malaysia in the midst of change, these changes in knowledge and practice contain a story about how the social terrain of the nation is being systematically reconfigured in accordance to precepts and assumptions singularly modernist in its constitution of reality and value. As a commentary on the "transformative" capacity of knowledge, its power comes from the ability of this knowledge to [re]-constitute the subjectivities of locals and their relations to self, to reality, to knowledge and to others. To the degree that this is so, the story is one about how the appropriation of certain discourses and the subsequent activities undertaken in the name of development—for example, the acquisition and deployment of quality knowledge—have contributed to the active reconstitution of local subjectivities and practices, whether individual or institutional.

Procedurally, this extension of a form of knowing and doing previously confined *outside* the social terrain of these targets is the *product* of the discursive heritage of modernity, whose power lies in its progressive ability to incorporate and transform social life (Escobar, 1994b, 1988) into forms amenable for treatment by its ruling institutions. In this instance, the discourse stipulates beneficiaries through the active production of categories of people for institutional or social *intervention* by looking at their education or training, terrain of engagement, use of time, organization of work, QCC participation, reliance on industrial standards and management capability. As noted throughout my preceding review of constructions forwarded by officials, and the regulatory or administrative forms effected and used by key organizing institutions (SIRIM, NPC, MITI), these are the very categories and descriptors relied on to identify subjects in need of training and to evaluate their efficiency [or effectiveness] as workers and institutions.

The progressive "un-layering" of interconnections—between knowledge and practice, constructions and assumptions, administration and aims, tools and rationality, institutions and regulation—undertaken in this chapter is a concerted attempt to pinpoint attention to those very practices and knowledge which make possible the very constitution and deployment of changes being considered here. A central organizing focus and procedure of institutional ethnography, the value attached to this attendance to the intersection between knowing and doing comes from its ability to make apparent *how* certain discourses, their claims, administrative or procedural links, work of ruling individuals and so on are successively [not to mention reductively] organizing and framing the everyday reality of people and institutions at the local level.

In the process, what is rendered open to scrutiny is the socially constructed nature and features of both the objects of intervention and the subjects of inquiry. Concurrently, by attending to the ways that knowledge and power interconnect, it is possible to see these objects and subjects not only as the products of discourses and practices being effected in the nation, but as *sites of production* where concepts, procedures and claims central to their own production are constantly being configured and [re]configured. Equally pertinent, this institutional ethnography of what is taking place and/or being put into place in Malaysia consistently bring our attention *back* to the implicit and explicit presence of a global Other who is centrally involved—economically, epistemically, functionally—in producing and organizing the individual, institutional and social terrain of the nation and its polity.

* * *

As I write, and these words show themselves to me across the screen of my computer, the tap tapping of the keys brings along with it a renewed sense of what was both gained and lost in the name of development. Linked here to a national sacrifice that had been wrought to equalize the social and economic terrain of the nation is the additional mediation and reconfiguration of activities and sight by the quality and market infused gaze of recent models of development. I wonder to myself, not for the first time, is the national sacrifice that is the NEP never coming to an end? Is it being offered up yet again for sacrifice as individuals seek to profit by it? Appropriating what is already a painful moniker in the history and memory of the nation in the service of seemingly non-national aims? Who are these individuals speaking as? Who are they speaking to? What indeed, are they speaking to?

While configuring the answers to these questions is partially why I am here writing this text, I know only that for now, both the researcher and returnee [I] are owners of a vocabulary that has been much expanded by words like "kiasu" and "bohsia."

Chapter Six

Mediating Culture and Mediation by Culture

This chapter considers the multiple ways that changes being put into place in Malaysia are themselves constituted by a process that simultaneously entails an active mediation *of* culture, as well as mediation *by* culture. In reference to "mediation of culture," I consider how these ostensibly non-local forms of knowing and doing are themselves effectively constituted and deployed in social, historical and economic terrains not their own. In this instance, the institutional deployment of QCC training and practices presented in the preceding chapter is only a partial component of this story on transfers. Embedded in this deployment process is also an active appropriation of certain local concepts by state planners, trainers and managers that *makes possible* the act of deployment itself.

In contrast, "mediation by culture" comes with and from the understanding that *neither* the locals involved in the transfer and deployment process *nor* the knowledge informing their locus of interest such as quality and management come empty of meanings. Given such an assumption, my primary interest lies in assessing *how* local culture may be a meaning-making filter and mediator constituting the active "translation" and experience of disseminators and recipient subjects alike in *their* encounters with these knowledge and practices.

It needs noting at this juncture that my concern here lies *not* in some effective identification of what the *definitive* Malaysian culture is. Rather, what I produce here is a recreation and notation of very particular encounters between the researcher "I/eye" and the people with whom I interacted that seems to speak particularly well to the question of "transferring knowledge in cross-cultural contexts." That is, whereas my intention in the previous chapter was to highlight the roles of these individuals as carriers of a westernizing and modernizing discourse mediated by and through the received practices and lenses of their respective professions, what I highlight here is their location as *conduits between* forms of knowing, given their location as owners of meaning systems, and the engagements effected between these sometimes synchronic and dystopic ways of understanding.

In the spirit of institutional ethnography, my intent is to render apparent some of the relations of knowledge and power, history and being, silence and voice, which may be organizing both their speech and the phenomenon of knowledge transfers. In the process, it may be possible to garner an alternative understanding of what "the local" itself may be about, culturally, historically and institutionally.

MEDIATING CULTURE: ON "HAV[ING] TO BE 'TWO'"

As I was to find over the course of my fieldwork, local managers and other like disseminators of knowledge have crucial roles to play as translators (e.g., Law, 1997a, 1997b) and mediators of non-local knowledge and practices. Central to this role is their ability to translate to the grassroots, and their ability to mediate between the two or more languages and worlds locating *them* at the cultural crossroads they inhabit. As noted by a manager at Infotech in describing this need to mediate between cultures:

> M: . . . for people like us, managers that interact more with the US, we have to be more like them . . . when they come . . . we have to relate to them in a more Western way . . . so we have to be "two" you know . . . then, we have to be Asian as well . . . and we must be able to translate this [Asian-ness] to a US context you know . . . so . . . managers that interact with the US have to be two-faced ah . . .
>
> V: [I nodded, looking at him. He pauses, catches my eye briefly.]
>
> M: . . . but not like . . . two-faced . . . in the way the Chinese thinks ah . . .
>
> V: [I nodded, smiling to let him know I understood what he was saying] . . . no . . . yes . . . I know what you mean . . .
>
> M: . . . some managers fail because they don't understand the cultural cues ah . . .

To him locals must, in their role as translators of the non-local to the local, have an understanding of the locality to which they belong, as well as being "conversant" in the discourse they are translating *from:* hence, the emphasis on "hav[ing] to be "two.'"

Trained by years spent acquiring their professional educations in the West, a point noted in chapter five, most of these managers and administrators are well-attuned to the precepts of their professions and to the nuances of living in the West, even as they are also carriers of a culture and a universe of meaning belonging to the specific intersections of history, memory, location and nation that are all their own. In this sense, they are crucial conduits mediating between a global knowledge and a local audience or context.

Descriptions such as the above also pinpoint attention to the mediatory space that these managers inhabit in their location as *local* managers working in non-local multinationals. It is a space that requires they have the capacity to take on or to "inhabit" a more Western persona when interacting with managers and the like from the US. In this regard, it is a process that seemingly requires that they relegate to the "background" their local Asian Self in such instances. Concurrently, in their official capacity *as* local managers, they are designated as representatives and interpreters of the local *by* and *for* the companies they are employed in.

As indicated in my earlier assessment of Malaysia's history (see chapter four), these representational possibilities are underwritten by the singularly modernist assumption that all locals are not only alike and equal in relation to the society they reside in, but culture itself is assumed to be always and constantly knowable and fixed, hence the possibility of relying on *specific* local representatives as the "eyes" to a *whole* nation, a people and a culture. By contrast, one of the points I seek to illuminate here is the difficulties of maintaining such representations since knowledge is mostly forged *in relation to* the specific site, interests, locations and so on inhabited by both the person speaking and the person listening. Knowledge and representation, in other words, are almost always situated (e.g., Haraway, 1988; Harding, 1998; Rich, 1984), and relationally or dialogically configured (e.g., Dwyer, 1987; Radhakrishnan, 1996, 1994).

As constituted under the auspices of these representational assumptions and expectations, local managers "have to be Asian as well" since it is *only in this capacity* that they are able to "translate this [Asian-ness] to a US context"; that is, by putting on, as it were, their official identity *as* local or Asian. This constitution recalls Dorinne Kondo's (1997) notion about the "performativity" of race and culture, albeit effected here in a somewhat different manner. Thus what is of paramount significance to the multinationals they work for are local managers' official location as Asian rather than any "living reality" as Asian(s). The existence of this representational role is underscored by the manager's use of "have to be"—as opposed to "are"—when describing the Self [*then, we have to be Asian as well*].

Frequently, this bridging by local managers often extends beyond the boundaries of translation to negotiating the vagaries and complexities of

Malaysian social life and bureaucracy. As exemplified by the following description offered by another manager from Infotech, familiarity with the local terrain is of paramount importance *for* his contribution to the company:

> M: . . . the building at the back [an addition currently under construction] . . . government regulations say 65 feet, our plan say 80 feet . . . so, I have to find outlah . . . uh, find out and see . . . meet people . . . talk to them . . . discuss . . . about what we can do . .
>
> V: . . . and?
>
> M: . . . we got 80.5 feet . . . [grinning as he says this, hands running lightly up and down his chest] . . . 80.5 feet . . . [chuckles a little before shaking his head] . . . so now I have to see what we can do, what we can do . . . *I don't think they can do that* [reference non-local managers' ability to negotiate the nitty-gritty of the local terrain, specifically, government bureaucracy] . . . the proposals, contracts . . . others just do the marketing, but *we have to follow* . . . [to] push through . . . make sure things get done after . . . agreements with the Malaysian government . . .

Local managers are thus crucial conduits *to the* local terrain both culturally and non-culturally, performing a bridging and mediation function that they seem to be more than aware of. As exemplified here, it is their ability to translate the Self to Others, along with active translations of Others to the Self, which gives symbolic currency to their presence *as* locals.

'Knowing" Culture and the Dialogical Location of Being "Two"

From the perspective of such an understanding, the focus on "having to be "two'" brings to our attention the implicit presence of a West that is constantly framing locals' work[ing] reality—whether in reference to organizational expectations regarding their efficacy to translate; the assumed representational capacity attached to their location as locals; or the implicit assumption of culture as fixed and knowable, possibly even "manageable"—while *also* seemingly silencing their presence *as* locals. Like the duality implied by the necessity to be "two," this seemingly inchoate series of expectations highlights the paradoxical location often inhabited by local managers.

However, that the manager quoted above is himself *not unaware* of the potential paradox or irony between this constitution of the self as cultured [*having to be Asian*], and the cultured-*ness* of the self as a living entity [*having to be "two"*] is evidenced by his need to qualify what he means by "two-faced" [*so managers that interact with the US have to be two-faced, but not like two-faced in the way the Chinese thinks*]. Of particular significance here is the articulation of this qualification *to me,* that is, someone whom he identifies as Chinese, albeit one also currently interacting with him as a researcher *from* the West, and how this possibility for articulation was negotiated.

Cued by my nod and our brief exchange of eye contact, he adds his qualification: in effect, a caveat stipulating that the reference to "two-faced" is in this instance, *less* a confirmation about the potentially negative constitution of this concept as manipulative or untrue than it is an acknowledgement about the reality of inhabiting and negotiating two languages and two [or more] cultures. The silent and cultured communication exchanged in that brief moment established not only *our* mutual understanding of how "the Chinese" [rather, the non-Western?] would understand the concept, it is also an implicit acknowledgement on *his* part, and affirmation of understanding on mine, of the danger that he might be *misread* by the [Western] *researcher I* who may [or will] see in his description a singularly Oriental trait associated with an inscrutable East more inclined than not to be dishonest and/or devious.

In this sense, the exchange between the manager and I is a commentary about the legacy and existence of an unequal knowledge and power relation underwriting histories between the West and the Rest (Said, 1979) which was, in this instance, implicitly *framing* and *organizing our* conversation *as* researcher and subject: an example of the living reality of having to be or being two. Equally of note, what was negotiated in the process of this exchange was a *transition* in our locations with respect to each other: one entailing the shift from researcher and subject to member(s) of a community [whatever or however this may be constituted]. This is significant because it highlights the dialogical and relational way that knowledge itself is negotiated into being.

From the standpoint of researching the cross-cultural, this recognition about the way that knowledge and relationships are constituted and produced requires on the part of the researcher, an awareness about the social situatedness or locations that people [including the self] are constantly inhabiting and interacting or negotiating from. This attention to the *shifting* feature of how identities and knowledge are mutually negotiated is particularly important in the context of constructing knowledge on/about Others because fine webs of history, knowledge and power are an endemic part of this negotiation. Put another way, understanding encounters between the researcher and subject in terms of such implicit and explicit issues of power and voice, history and location, entails a recognition by the researcher that his/her engagement with subjects is fundamentally one marked by the respective locations of each *as* knower and known, and whatever this may imply about the practices and relations of knowledge and power that is at play.

In the absence of such attentiveness to location and voice, the local other stands vulnerable to the power of the researcher self [or any other knower for that matter] to "know" or to misinterpret and so, to silence. It is for these reasons that the methodology that I rely on and use in my analysis above entails an active attendance to what people have to say, what they do say, and how they say it *as ways of getting-at* how culture "works," as opposed to *what* culture is.

Accordingly, this attentiveness may include an awareness of how relations informing or underwriting encounters between the researcher and his/her subject(s) may be part of what is organizing what can be said and what is/is not said within such encounters. For example, the manager telling me about the "having to be two" space he inhabits feels the need to qualify where he is coming from because he is aware that he may be misread culturally in the absence of this qualification. His awareness and knowledge about this potential occurrence, along with the [implicit] relation of power and knowledge constituting our exchange, are what is framing and compelling his need to qualify. What is [silently] written into and anchoring this tableau is a history and an awareness of how knowledge on [and about] the subject-Self may be constituted by Others who are researchers from the West.

The task of the researcher in such instances is *not* to document the manager's description as if it was representing who or how *he* is. Rather, the task of the researcher is to attend to what might be organizing his speech or description to begin with *so that* both what is of import to him, and what he is contending with could be known and rendered visible. In line with institutional ethnography's aim of "leaving subjects alone," this attentiveness is a process of [negotiated] understanding involving active attendance to how subjects see or represent themselves, their relation to others, to the knowledge and practices on hand and so forth. In effect, rather than any *prima facie* attempt at identifying or categorizing them in accordance to the predefined schemata, presumptions and so on that the researcher may have about them, the researcher attends *instead* to whatever it is—history, discourses, culture, location—that is organizing their speech and actions, local and global. In this particular instance, the effectiveness of the manager *as a knower* and a mediator between languages and cultures, locations and voice is evinced by *his* attentiveness to the multiple forms of "nuances" and relations that are organizing the engagement between him and I.

"The Malaysian Way"

As an example of how the local may be articulating itself within the non-local, the following description of QCC practices offered by a local quality promoter and trainer provides a confirmation to observations detailed above, while also signifying in itself, the way[s] by which the local "makes itself known." Specifically, the words forwarded here not only show how local culture filters and mediates the apprehension and meaning of the trainer whose words they are; at some point there is *a shift* in his location from "[just] trainer" to "Malay[sian] trainer":

> T: . . . the Japanese are quite strict about the numbers in the [quality control] circle, they restrict it to seven at most, *for Malaysia, we have to be a little more flexible,* circles may be as large as 15, numbers is not an issue, although we do try and find out how they allocate work amongst themselves. . . . Another difference may be the duration of projects, the Japanese are well-known for their precision in organizing things, usually six months, *for us, we have to be a little more flexible,* some may be a year, may be more . . . *this is the Malaysian way* . . . the Japanese go straight to the point, *Malaysians tend to have fancy, fancy stuff* . . . [gesticulating his arms a little as he goes on] . . . *they have music, special clothes or uniforms, the kebarung is a craze now* . . . anyway, for the presentation, the Japanese go straight to the point, maybe they need only 12 minutes, *Malaysians maybe 20 minutes . . . that's 8 minutes extra!* [gesticulates with his arms again. I chuckle.] . . . what to do with this? . . . *they doodle, enjoy themselves, some have not only music, but pantun* [poetry] . . .
>
> V: . . . yeah, I've seen some of them, I think it's great!
>
> T: . . . yes! [chuckling as he continues] . . . there was even one group . . . of Indonesians I think . . . who started with *silat!* [Malay form of self-defense somewhat like taichi; he shakes his head as he continues to grin] . . . the Japanese did not like it . . . it drives them crazy . . . they think it's like selling medicine in the night market!
>
> V: [chuckle] . . . n-no . . . I think it's rather nice actually . . .
>
> T: . . . *so, rather than black and white, we have color* . . . some of the material is rather dry, so they make it colorful . . .
>
> V: . . . yeah . . . I think it's wonderful [nodding again]
>
> T: . . . also, during presentations, Malaysians usually have more than one presenter [I nod] . . . the Japanese normally have only one, Malaysians have two or three . . .

What is instanced above is an active "showcasing" of differences between self and other. While QCCs are highly standardized and regulated as a procedure and form of analysis, the meanings and forms attached by locals *to* QCCs as an actual practice is quite localized in its constitution.

Underlined by an emphasis on the need "to be flexible," this localization is not just a matter of process or meaning. It filters through to affect the actual structuring and form of these circles. Constituted under the auspices of flexibility and "culture" [*this is the Malaysian way*], there is an almost wholesale reconfiguration of what QCCs should be like. As noted by the trainer, this is often a reconfiguration that does not "sit well" with expatriate managers from multinationals located in the nation.

Equally of interest is the manner and speech of the trainer conveying this information and knowledge. Starting with what seems to be information of a more objective nature regarding team size, duration of projects, length of presentations and so on, his demeanor changes in form and tone as he goes on to elaborate what QCC amounts to in the local context. Reinforced partially by my own enthusiasm and interest in what he was saying, he seems to literally "loosen-up" in behavior, evinced here by the gesticulations and movements of his arms, his [rather, *our*] recurring bouts of chuckles and so forth. He seemed to "become" a part of the experiences and social nuances he himself was describing; hence, my earlier observation regarding the shift in location that occurred in the midst of his description.

As in my experience with the manager from Infotech, there was also a shift in the relation organizing both our locations, moving from researcher and subject to a seeming commonality as "Malaysians" [aware of local forms, history, preferences] and "professional insiders" [trained and well-versed in the language and practices of the management sciences]. This "bounding" of identity was configured via a shared understanding—unvoiced and not fully articulated—of the implications attached to what he was describing [local QCC practices] *on* received procedures of control and regulation, not to mention expectations, that exists in modern corporations. It was a silent knowing established and confirmed by *his very description* of how local QCC practices are a source of "irritation" to certain managers from elsewhere; that is, exaggerated, injected with humor, relying on analogies seemingly a critique of the self but which comes at the expense of the subjects under consideration. Also contributing to this renegotiation in location and knowing was an implicit criticism on his part [and understanding or agreement on mine] over the seeming insensitivity of expatriate managers, and inflexibility of received organizational approaches: critiques reinforced here by what amounts to his summary statement, *"rather than black and white, we have color."*

Presenting and Translating the Self to Others

A point to note here in relation to knowing *across* cultures is the role that these managers and officials play in presenting and translating to the researcher I, the specific features of the locality in which we are embedded. As exemplified by the exchange above, the trainer was also actively constituting and delineating for me the particular features and cues to local culture that I, the researcher, should be attentive to. Indeed, these instances of cuing and active presentations of the local serve as a reminder of sorts about the "having to be two" space these individuals inhabit.

The following is another such example. In this case, the director of one of the skills development centers [a joint venture between industry and government] I visited was actively engaged in conveying a sense of what collaborations like this joint venture was about in "Malaysian terms":

> D: . . . this "management by collaboration" is very Asian . . . not like the Western . . . self-centered and individualist . . . there's a Malay word "*gotong royong*" . . . that's the spirit of our culture . . . it is a community effort . . . you know how Malays build a house in the *kampung?* . . . it is based on cooperation from the community . . . the government and private sector sitting down together and working for solutions . . .
>
> V: . . . so is this the new version of gotong royong?
>
> D: . . . so yes . . . this culture of community is part of our style, an important aspect of management style . . .

Not only does the director introduce the concept of "gotong royong" to me in his bid to situate the meaning of "Malaysia Inc.," he goes on to elaborate what the features and "genealogy" of this collaboration entails. As he does so, he concertedly opposes the Self *against* an Other [the West] in order to lay claim to his assertions that this form of management is very Asian.

As denoted by these examples, local managers and officials are constantly involved in translating and presenting one form of knowing into another, whether non-local to local, or local to non-local. By attending to the cues forwarded by them, not only am I privy to the potential ways the local may be articulating itself (e.g., QCC practice, industry-government collaborations), it is also possible to garner a sense of the constitutions, meanings and locations inhabited by these individuals *as* members and representatives of the local: that is, how they see, what is important to them and so on. In the process, not only is knowledge gleaned about how they may look to us, or what is important to them, but perhaps more important still, it is equally possible to begin seeing how *we* [as the Other] may look *to them.*

LOCATING CULTURE AND LOCALIZATION: THE CONCEPT AND PRACTICE OF GOTONG ROYONG

In this segment, I focus on the multiple mediations of and by culture that entered into the constitution and deployment of quality know how in Malaysia. Particular attention is trained on the local concept of "gotong royong" and its various articulations in and throughout this process of constitution and deployment. Often associated with Malay rural life and practice, encoded in the concept is a belief about the interconnected nature of how and what things are like, whether in reference to relations of people to their community, their engagements with each other or their relations to the larger social, economic and physical environment. Used in this way by locals to represent their relation to each other or to the social community, it is often articulated in terms of actual and concrete actions like community clean-ups, house-moving, the construction of community-oriented buildings and so on.

The symbolic currency attached to the concept came to my attention early on in my fieldwork. During one of the first TQM communication sessions I attended at Multitech, the champion [team promoter] for one of the groups kept reverting to the concept in order to emphasize the centrality of participation for TQM practice. According to this trainer-translator, "*participation ini semacam prinsip gotong royong ah gotong royong . . . tahu tak?*" Initially somewhat taken aback by this allusion to a concept I had long relegated to the realm of memory, this active identification and association of gotong royong to the exigencies of TQM in practice was one I would encounter over and over again in one way or another while in the field.

From the standpoint of those officials and trainers vested with this responsibility of deploying TQM knowledge, gotong royong was a crucial way for communicating and translating the central beliefs and assumptions attached to this otherwise non-local form of knowing and doing. In effect, the concept was actively appropriated and used by these individuals as a cultural filter for making meaningful the otherwise foreign essence of teamwork as constituted by modern quality discourses. As noted by a senior staff manager at Infotech, there is always a need to "tailor" the objectives and activities of the company to the locality:

> . . . if you just leave it as a global thing . . . people can't relate, you see . . . the guy at the grassroots must be able to relate to it.

As he goes on to note in his description of Infotech's "way of doing things," it is only after such tailoring of "[the company's] culture" that locals working for the organization "can relate to it" and figure out "what they can all do on site."

As exemplified by the words of this manager, what seems to be needed when transferring or applying know how from one locale or culture for use and adoption elsewhere, is the issue of *how to "localize" it.* Or, as noted alternately by a productivity and quality promotions officer and trainer from the NPC, they "try to put it in the Malaysian scene."

On a personal and social front, this resurgence and revival of the concept of gotong royong was particularly interesting to me. It had been a concept regularly used by politicians and officials in "naming" (Mueller, 1987b) and explaining their activities [especially around election time] throughout the socially and economically fragmented post-Emergency years of the 1970s. It was used to underline and remind all communities about the interconnectedness of the social, physical and even economic world we lived in. During this time, teachers and schools intent on encouraging students to participate in a variety of clean-up efforts [of the classroom, the school compound, the surrounding neighborhood] would also call on the concept to cultivate this desired outcome. In this sense, gotong royong was the practical equivalent to classroom teachings about civic responsibility and duty. The concept was less prominently paraded around, at least in the public arena, throughout the 1980s until this re-emergence in the 1990s.

Gotong Royong and the Practice of Teamwork

This cultural mediation of QCC activity by and through the concept of gotong royong is however, not just a simple act or process of translation. This culturally mediated deployment *changes,* in actual meaning and form, the very practice of these activities. Specifically, through its constitution as gotong royong, the concept and practice of teamwork is infused by a whole series of local associations and meanings that *precedes* local workers' acceptance and/or subsequent application of its precepts and claims into practice. It is an infusion that colors the very process, meaning and feature of what participation is as a workplace activity.

Of particular significance here is the community-oriented feature of the concept and its association with local practices of collaboration often to be found in the local community. As an example, in the various QCC presentations I observed, most if not all involved a good proportion of team members who actually did the presenting. For those not involved in the actual presentation, team members participated in other capacities in support of the presentation taking place: coordinating overheads, holding up posters, distributing flyers, highlighting points on overheads.

In the specific case of a QCC presentation I observed at the regional QCC convention organized by the NPC, two [female] of the five members were designated as "speakers," while the other three [male] were responsible for attending to the transparencies and slides. As the two women presented, appropriate transparencies would come on. Occasionally, the women would alternate in their presentation; sometimes using a questioner-responder format where one would ask the other a question [usually for clarification], which the latter will subsequently explain. This seems to help make the presentation of their material clearer to the audience. In general, the whole presentation—text, transparencies and slides—were synchronized down to the very last detail, making for an extremely smooth presentation.

Another particularly memorable group included a member who introduced his colleagues and their project by way of a *pantun* [poem in Malay], whose imagery provoked chuckles from members and audience alike as he went around and poked gentle fun at all members' individual idiosyncrasies, and their scramble through the project. I make the reference here to the use of this cultural form as a mode of expression because poetry in the context of Malay social history and culture is often used not only as a form of speech. Much like the oral lyric poetry of Bedouins in western Egypt studied by Lila Abu-Lughod (1986), inherent within the form and structure of Malay poems, not to mention strategic use of images, language and words, is a tradition of irony and social criticism. Best exemplified by the tradition of *dondang sayang,* implicit in this performance of dialogical exchanges between participants, usually two, and usually consisting of some repartee between men and women, are often very biting social critiques or commentaries on a variety of subjects, including gender relations and other social or political issues.

I remember well the enjoyment my father, who was a big fan of the genre, would glean from watching such performances, his laughter emerging often as the repartee exchanged between the protagonist and antagonist increased in "bite." As an example of such an exchange, a woman might, under the guise of an exaggerated coyness, itself an implicit critique of socialized and patriarchal notions of womanhood, launch a volley of shots at her male counterpart, all the while satirizing and poking fun at him and/or the variety of male attributes, privileges and so forth that he is taken to represent. Quick in exchange and sharp in tongue, not only is the genre testament to the quick-wittedness, erudition and keen observation of the participants, it is also a form of expression rich in humor whose value is derived from participants' ability to play with and play on entrenched social images and practices via the use of language and words. While I do not have enough knowledge of the scope, frequency, instances and forms of poetry-use being expressed by QCC members in their presentations to be able to assess the implicit or explicit commentaries which such articulations may be voicing, the *very* association, use and abstraction of the form to the context of quality know how and practice is an interesting notation on how TQM [and QCC participation] *may be* experienced and constituted by those over whom it is deployed.[1]

What is illustrated by the examples above is not only a commentary about the way local forms of cultural performance or expression can filter through to an ostensibly non-local form, they are also instances of the collaborative and participatory spirit of *gotong royong in practice.* In these examples, participation *extends beyond* the actual conduct of work that team members perform *behind* the scenes to include a public claiming of membership, identity and accountability.

Of particular significance to this claiming or show of membership is the garb that members of teams wear during their formal presentations, whether in the company or at a convention. There is an almost uniform-like consistency to the clothing of team members. Most of the men are inevitably wearing guayabera-like shirts, the women kebarungs [or sometimes dresses if they are non-Malay], made from the same fabric, usually local batik of one kind or another. Such "performances of identity" (Kondo, 1997), and symbolic representations of members *as a team* construct for those observing them an indelible image of these individuals as a unit. Moreover, the use of local batiks and the choice of kebarungs and shirts by members index yet another level of identity: namely, their location as Malaysians. While this may be constituted as a representation of the Malay identity, the use of batiks and [at times kebarungs] by non-Malays seems to contradict such an observation.

What is significant from the standpoint of representing [or identifying] Others is the potentially *homogenizing* and *misleading* constitutions that can be made as a result of the meaning and location that are attached to what we normally assume to be "obvious" denotations of culture, such as dress-form or nationality. In this case, the use of batik and kebarungs by both Malays and non-Malays *blurs* the boundary between ethnic categories, rendering attempts to locate Others in accordance to these demarcations risky at best. Concurrently, the location of batik and kebarung as indices of both ethnicity *and* nationality render potentially problematic any easy constitution of Others in accordance to either. This is particularly so in the absence of a more situated knowledge about the specific locations or intersections of location inhabited by individuals making-up our objects and subjects of inquiry. This potential to homogenize or essentialize Others is doubly problematic in the context of nations like Malaysia where lines of demarcation such as race, ethnicity, class, gender and religion are constantly intersecting and bisecting one another, making for an ever-shifting social configuration which can only ever be known *in relation to* the social particularities which are organizing the specific event, interaction and so on that is under observation.

Of equal significance is the implicit commentary and juxtaposition in constitutions of self and meaning that are attached to this locally mediated practice of teamwork and participation. In contrast to the universalizing and economizing constitutions of subjects forwarded by official discourses on quality, the focus on community, collaboration and claiming of identity observed here call into question all reductive constructions. As an example, Multitech's metaphor of quality control as a chain made up of links silences the possibility of seeing the potential meanings which may be attached to the experience of participation itself *since* participants and the meaning of participation come *already known,* and *de-linked* from any notion of Others as socially, culturally and historically located beings.

By contrast, the story I tell here focuses our attention precisely on the existence of this social, cultural and historical situatedness of Others, and how it mediates their engagement(s) with ostensibly universal forms and practices. In particular, the active appropriation of quality and teamwork to the exigencies of national development *and* gotong royong, as well as the constitutions of self in these terms, call into question the very possibility of attending to transfers as a simple focus on work practices or application of tools and know how.

Gotong Royong and "Malaysia Inc."

Concurrently, the concept of gotong royong also mediates how changes in relations within and between institutions and/or sectors are understood and effected throughout the statutory, government and industrial sectors. As forwarded by the director of the skills development center quoted earlier, joint efforts such as "the government and private sector sitting down together and working for solutions" are themselves an instance of the collaborative spirit of gotong royong in practice.

Constructed under this localized eye, alliances between government and industry, both local and global, are often culled and deployed under the banner of development itself or the social responsibility of industry *to* this development. As noted by the aforementioned director:

> . . . our objective is to look at a place which can spearhead human resource development for society, not just technological but managerial as well, *which can help* or *are needed* to propel development in the country.

Commentary on the infiltration of modernist values in the above notwithstanding, what is constituted here is the need to cultivate efforts or initiatives that are *in the interest* of national development. Or, to re-quote Mahathir Mohamad himself:

> . . . . [i]f they are not brought into the mainstream, if their potentials are not fully developed, if they are allowed to be a millstone around the national neck, then *our progress is going to be retarded* by that much. No nation can achieve full progress with only half of its human resources harnessed (1993b: 415–416, emphasis mine).

Not only is there a reliance on the community to participate in the nation's development, this reliance is already embedded in a notion of development which is community and nationally oriented in nature and form. It is an embeddedness which is understood more often than not by locals as an instance of gotong royong in practice.

Seen from this perspective, the concept of cross-sector collaboration [signified by "Malaysia Inc."] becomes in effect, a way for the government to *ensure* that it is the well being and development of the *nation* and *community*—in particular, the Bumiputera community—that is at the center of attention [as opposed to deregulation or privatization for example]. Development, in other words, will be based on ties and relations of collaboration bounded *not* to the exigencies of free-market enterprise and deregulation, but to national interest and Bumiputera development. Hence, my earlier observation that Malaysia Inc. is used actively by state planners to compel the participation of industry *in the service* of the nation's interests.

The concept of "Malaysia Inc."—barometer of what multilateral agencies like the World Bank and the International Monetary Fund (IMF) see as a symbol of the nation's move to deregulate and privatize—is thus actively used by these officials to call for other forms of collaboration. What is reconstituted in the process is the very notion and meaning of collaboration itself, in particular, industry-government collaboration.

Local officials and state planners are seemingly abiding in this regard, by a concept of collaboration that is more broad-based and community-oriented than that normally constituted by various discourses of collaboration to be found in the West, such as collaboration in deregulatory or joint ventures. As exemplified by another quote from Mahathir Mohamad, "Malaysia cannot deregulate if . . . the freedom afforded to enterprise becomes merely the freedom to exploit *without any sense of social responsibility*" (1993b: 419, emphasis mine), discourses of deregulation and free market agency are actively *reconstituted* by the state under the auspices of "obligation" and "social responsibility." In the process, Malaysia Inc. as a symbol of neoliberal policy is actively *recast* and used by state planners as a discursive entry-point for forwarding their *own* aims with respect to Bumiputera and human resource development, economic defense and the like.

Equally important to note at this juncture is a caveat to the story I am actively *re*-presenting here: namely, the analysis I forward above should *not* be taken as some *essentialist* commentary attached to the local[ized] notion or practice of collaboration [for example, gotong royong] that exists in Malaysia. Rather, what I am documenting is the *cultured-ness* of engagements, meanings, understandings and so on—and the embedding of these in situations of uneven power and voice, inclusion and exclusion—from wherein knowledge, practices and institutional links are forged, both by a knowing Self [researcher, multinationals, industrialized West], and an *equally* knowing Other [local officials, managers, state planners].

To this end, my re-telling of this story is more a comment on *how* gotong royong as a concept and experience to which most locals are privy is currently being *re*surrected and *re*constituted *in relation to* very specific interests and aims, both local and global. The need to render the non-local understandable to the local, the tension between national interests and global aims, local meanings and global precepts, community and market, social responsibility and deregulation and so on are all part of the plethora of knowledge and power relations organizing and framing this resurrection and reconstitution of the concept. Put another way, within the situated and organized terrain informing and compelling the nation's adoption of non-local knowledge, gotong royong is a way to mediate between the conflicts and paradoxes engendered by having to negotiate between local interests and global pressures, pre-existing meanings and changing identities, local notions of self and modern requirements for participation.

Gotong Royong and Economic Defense

In connection to officials' interest in this community-oriented collaboration and the reconstitutions in meaning and forms [development, Malaysia Inc.] observed thus far is the need to marshal the economic defense of the nation, which I called attention to in chapter four. To this end, visibilities regarding power and voice, inclusion and exclusion opened up to scrutiny by the prime minister (Mohamad, 1993a, 1993b) become effective social forces organizing officials' [and locals'] subsequent constitution of development efforts. What struck me as I traced my way through the myriad of public and semi-public institutions in the nation was the degree to which concerns and activities at these levels were themselves articulated and re-articulated in terms reminiscent of the prime minister's concern with issues of Malay backwardness and defensive necessity. As in the very formulation of the strategy that led to the activities and changes being presented here, local concerns and understanding constantly color not only what is being put into place, but *how* it is being put into place.

Within this reconstituted framework, interest in training workers is no longer a simple question about producing economically relevant subjects. According to a top official from the NPC, a reason why the government is "trying to force the SMIs" to participate is:

> . . . we cannot hope for the MNCs to be here all the time . . . they have their own training . . . skilled workers . . . but maybe they will go . . . if competition increases and labor costs go up . . .

As he goes on to add, such efforts are important for the country since MNC activities account for only 15% of economic activity, whereas 50 to 60% of it comes from small and mid-sized industries. Similarly, regulatory requirements about quality improvements, training and research and development encoded in the various ITAF schemes become ways of ensuring that local SMIs have the wherewithal to compete in the international marketplace *without* losing sight of the community interests informing such requirements to begin with.

This is best exemplified by the following reasons given by an official from the Ministry of International Trade and Industry (MITI) in explaining the need to regulate local participation in trade. According to him, all local companies have to enter into an approval with MITI before they enter into a contract with foreign parties because:

> . . . most Malaysian companies are young, they normally don't have much experience . . . the government thinks that at some point the government should protect the company . . . we try to ensure that the royalty paid to the outside party is commensurate to the level of technology transfer . . . in screening these agreements, we impose the condition that the governing law is the governing law . . . [and that] the settling of dispute must be in KL . . . we have had experiences in this line in the early years where contracts say that only the licensor can have the right to terminate the agreement . . .

Construction of local SMIs notwithstanding, regulation itself is a way of protecting the interests of local businesses and industries: that is, a way of confronting the inequality in relations, experience, expertise, control over technology and so forth indexing relations between the West and the Rest.

What is interesting to note here is the effective appropriation of the "legalese" embedded in discourses of international trade and foreign direct investments, and the active *reflection back* of this legalese to an industrialized West, formulated here under the guise of a legislated requirement and mandate of approval by state agencies prior to locals' entry into the fray.

Or as constituted by another government official, this need to build-up the defensive capacity and immunity of the locality is necessary because of the historical domination of developing nations by the economic and technological might of an industrialized West. Speaking in reference to the need to increase local proficiency in the areas of science and industry, this Ministry official observes that while Malaysia "ride on MNCs," there has been:

> . . . little or no actual transfers of technology. So if we want our own, how do we generate it? . . . this can only be done through the local companies . . . through encouraging students to go into the sciences . . .

Written-into these words [and actions] is a concerted effort to search for and negotiate challenges and solutions that will be in the interest of the nation. But as shown in the examples above, solutions being articulated also entail the active appropriation of those very knowledge and forms relied on by the industrialized West in its formulation of strategies. Thus, much like the paradox inherent in Mahathir Mohamad's appropriation of neoliberal development policy noted previously, these examples are equally paradoxical and contradictory in their constitutions of the challenges and solutions available. On the other hand, however paradoxical and contradictory, these constitutions are consistently negotiated and/or deployed *around* the interest in collaboration that I call attention to here: specifically, collaboration *by* community *for* community in the name of national economic defense and Bumiputera development.

In sum, while many of the programs and objects of attention here, like quality control, and research and development are typically constituted under neoliberalism as crucial components of competitiveness and efficiency; in the case of Malaysia, the meanings and reasons underwriting their constitution and adoption are singularly localized and mediated by local concerns like gotong royong, national interest, social responsibility and defensive necessity.

HAVING TO BE "MANY": *GOTONG ROYONG* AS ALTERNATIVE CONSTITUTION OF DEVELOPMENT

Throughout this discussion it has been my aim to show how gotong royong has become a broad-based notion of collaboration in Malaysia. Framed by another ostensibly Malay, albeit Islamic notion of what collaboration entails—"obligation"—the concept has gained new currency in the context of local development. The existence of this potentially broader constitution of collaboration came to my attention during a conversation with one of the managers at Infotech regarding the kinds of linkages (e.g., contractual, collaborative) that exist between the company and the state. In the following excerpt from this discussion, the manager pauses quite abruptly and changes his subject-verb combination [from "I" to a seemingly *non*-Infotech "we"] in the midst of his description; a change [in location] which seems to color what is to follow:

> M: I have a goal . . . my . . . [stops somewhat abruptly, pausing a little] . . . *we* [emphasis here. I wondered to myself as he said this, who he meant by "we"] . . . *would like to see one percent* [emphasis] *of Infotech Malaysia sales . . . back in the country* [holds up the index finger of his right hand, gesticulates a little, before turning and hitting it against the table-top in emphasis as he said this. He pauses as I look at him, nodding.] . . . do you know how much Infotech made last year in Malaysia alone? [I shake my head] . . . over two billion dollars [emphasis] . . . two billion dollars each year . . . in [emphasis] . . . US dollars . . .
>
> V: I know, it's incredible [shaking my head] . . .
>
> M: What is it? . . . what is one percent of two billion dollars? [brief silence as we both tried to comprehend what this translates to]
>
> V: One percent . . . of . . . two billion [drawing imaginary figures on my notebook] . . . twenty . . . million? [I looked at him inquiringly as I tentatively offered this information]
>
> M: [nods his head] . . . that's right . . . twenty million [pause]. Do you know how much that is? Right now, we [*note change in location here back to "role" as manager*] spend about two million each year . . . and we have to work hard to justify this . . .
>
> V: I don't suppose headquarters will just say, "here, okay—take it" huh?
>
> M: [he chuckles, hanging his head a little over the desk and shaking it]
>
> V: No? I didn't think so . . .
>
> [Detour in conversation here to some other event. After a brief pause in the midst of this, he asked me where we had gotten to in his list of "to dos']
>
> M: Let's see, u-uhm, we have the Yayasan Bumiputera [In Malay, "Yayasan" means Foundation] in [Malaysian state] . . . and [pause] . . . do you know what "poverty" is called in Malay? . . . know or not? [looks at me briefly; I shake my head] . . . *basmi kemiskinan* ['basmi" means eradication; "kemiskinan" poverty] . . . Yayasan Basmi Kemiskinan [another

> Malaysian state] . . . the Yayasan Bumiputera provides scholarships to Bumiputeras. . . . We have one week workshops for poor children. . . . Most of these children, they come from poor families . . . living in kampungs . . . we bring them in, give them an orientation . . . introduce them to the universities and factories [pause] . . . we bring their parents in for a half-day seminar. . . . Hopefully they will see the light. . . . Hopefully . . .

As exemplified by the above, close connections exist between global capital [*Infotech*], development [*Malaysia, Bumiputera development, poverty*] and poverty eradication [*reinvestments, aid, scholarships, workshops*].

Underscored by his change in location and identity, the manager launches what amounts to an acerbic critique and indictment of global capital and the massive amounts of wealth accumulated by institutions like Infotech in Malaysia with little regard for [re]-investing or [more generously] sharing such wealth *with the country* from whence it is being generated. Shifting his location back to that of "manager" [as opposed to Malaysian, Malay and so on], he double-verifies the "authenticity" of his claims by speaking as an "insider." *In* this location as manager, he confirms how hard it is to get the company to reinvest or share its wealth back into or with the nation.

Framed by the discussion on collaboration(s) in our conversation, his description speaks of an attempt to reconstruct for the researcher I the specific linkages between the company, development and poverty eradication. Embedded in his description is also an attempt to *re*-articulate the possibilities for an *alternative* development through a form of collaboration that does not now seem to exist; a form of collaboration seemingly based on a more broad-scale and localized notion about what "aid" and "poverty eradication" entail.

Of particular note here is the juxtaposition between his critique [and silent valorization of the need for more reinvestment and sharing] and his specific reference to what poverty is in Malay [*do you know what "poverty" is called in Malay? basmi kemiskinan*]. While the subsequent description is more a definitional clarification [not to mention, opportunity to witness the modernist constructions of underdevelopment embedded in his constitution of both Malays and poverty], the connection between what had come before [his critique and implied existence of an alternative], and what is at the current focus of discussion [Malay constitution of poverty] "cues" attention to the possibility that his *previous* critique may have been mediated by constitutions, assumptions, beliefs and so on *extending far beyond* a mere discussion or critique of global capital and development on the one hand, and his official designation as manager or even nationalist on the other.

What is significant to note in light of this *multiplicity* is the equally multiple locations the manager traverses over the course of his description, along with the changes in constitutions and visibility attended to in these locations. The logic of his discourse—organized by the presence of multinationals, the vast amounts of profit they make in the nation, and the lack of sharing which is the object of his censure—*changes* as his very *relation to* this object of censure appears to shift along with his subject-positions.

Modernist values and assumptions aside, what is interesting is how the manager [explicitly or implicitly] articulates alternative possibilities *only* when speaking as either "Malaysian" or "Malay." In the first instance, reconstitution seems to lie in greater reinvestment or sharing of profits back in the nation. Under the guise of the latter, connected with his focus on Malay and Bumiputera development, is a reconstitution of what collaboration is [poverty eradication, aid for the community, sensitivity to the plight of Bumiputera poor], and/or how this may be effected [scholarships, orientations].

It is worth noting that what may be framing this manager's specific constitution of collaboration [and development] is a potentially fine web of meanings associated with his location as a practitioner of Islam. Specifically, I refer here to a mediatory possibility traceable to the five Islamic precepts that all Muslims are supposed to abide by throughout their lives. A series of obligations to be guided by, included within these precepts is a specific call to believers in the faith to be "charitable to the poor." As a reminder that development may possibly be as much an issue of "morality" and "spiritual responsibility" [or obligation] as it is efficiency and competitiveness, his speech [as a Malay and a Muslim] may well also be an attempt on the manager's part to articulate a form of "help" that is more [if not totally] unconditional, and founded on [an implied privileging of] empathy over pity, morality over materiality and spiritual obligation over economic interest.

While I do not, and cannot firmly claim that the manager's critique and attempt to articulate an alternative form of engagement between global corporations and development is directly connected to his religious beliefs, the constitutions and associations embedded in his changes of location pinpoint attention to the likelihood that there is "something" *other than* profession, class or even nationality framing his speech.

What the foregoing story highlights is the presence of a locality and a local knowing or being marked by a multiplicity of locations. As in the case of the manager from Infotech and the "having to be two" space that he inhabits, the manager above is equally constituted by a plethora of locations that are constantly *shifting* depending on what is under discussion, who he is interacting with, what he is speaking about, and so forth. In contemporary Malaysia, where locations are already marked by the pluralistic heritage of the nation, the so-called entry and subsequent "infiltration" of ever more modern and global forms of knowledge and practices into the locality are only adding to this already heterogeneous terrain. Thus constituted by these local and global simultaneities and locations, individuals are required to constantly mediate between and within the shifting boundaries of the location [and sites] they find themselves in. It is a simultaneity and heterogeneity that can, on occasion, lead to a sometime paradoxical and dystopic multiplicity.

From the standpoint of how we "know" Others, this attendance to location and voice questions the assumption of modern nations as some unified terrain, closed off from the exigencies of the global terrain surrounding it, or the multiple pluralities and diversity constituting it. By extension, awareness about this multiplicity making up "the local" calls into question a notion of culture as knowable, fixed and representative of some homogenized knowing that can be and is reducible to the level of the nation-state, so much so that "cultures" and "nations" are often taken to be one and the same. By contrast, what I forward here is a story about the multiple, ever-shifting and socially-constituted locations inhabited by locals [and nation alike] whose very location can only ever be known in relation to the specific articulations, exigencies and so on that are constituting and organizing them and their visibilities.

Following from the above, we find in the words of Mahathir Mohamad below, a constitution of the self that simultaneously echoes and produces the sometime paradoxical multiplicity which is contemporary Malaysia today:

> Malaysia should not be developed only in the economic sense. It must be a nation that is fully developed along all the dimensions—economically, politically, socially, spiritually, psychologically and culturally. . . . There can be no fully developed Malaysia until we have finally overcome . . . the challenge of establishing *a united Malaysian nation* with a sense of *common* and *shared* destiny. . . . [P]ractising a form of mature consensual, community-oriented Malaysian democracy. [And] establishing a fully moral and ethical society, . . . [where] . . . Malaysians of all colours and creeds are free to practice and profess their customs, cultures and religious beliefs, . . . a fully caring society and a caring culture, . . . [where] society will come before self, . . . [and] a society in which there is a fair and equitable distribution of wealth of the nation, in which there is full partnership in economic progress (1993b: 404–405, emphasis mine).

While references to the community oriented and socially responsible features of his vision recalls the embedding of local actions and meanings in the concept and practice of gotong royong, the prime minister is also—in his bid to constitute for the nation, a Unified Self—adhering to a constitution of the locality, and a notion of the nation state that is singularly modernist in its assumption about the possibility of having or being "one" [cultural, location-wise and so on].

In this instance, the search for a development "without being a duplicate of" (Mohamad, 1993b: 403) the West sits in paradoxical tension against the simultaneous requirement and assumption that "unity" is possible. Not only does the logic of his analysis require an effective *de-linking* and *abstraction* of polity and nation *away* from the pluralist heritage and diversity that is the nation's history and living reality, it is also a model of development that calls for and requires that this same polity be "owners" of multiple forms of "language"—modern, professional, economically-relevant, scientific—whose very relevancy and value as contributors to the nation's development is premised on the ability and capacity to mediate between and within these multiple forms of knowing, doing and being. There is a fundamental paradox and contradiction encoded in his presentation of an alternative development and national possibility since this focus on multiple capacities and multiple values is *precisely* what is *not* allowing for the possibility of having "one self" or "one nation."

CONCLUDING REMARKS

The specific constitutions and possible mediations noted in this chapter highlight the localized meanings and multiplicity of intentions that are framing current changes introduced in Malaysia. "Culture" as evinced by the stories presented here articulates itself in multiple ways. In the case of transfers, the presence of culture—rather, the cultured-ness of what is under consideration—manifests itself via the need by managers and trainers to actively mediate and translate between the various forms of knowing and doing that they are constantly negotiating in or with.

In reference to QCC practice, not only is this know how translated and deployed according to very local concepts like gotong royong, the subsequent adoption and use of this know how is localized *in accordance to* the precepts, assumptions, meanings and so on attached to *these* concepts as ways of knowing and doing. Thus we find that quality in Malaysia is not just about the active linking of individuals to organizational aims, it is about being members of teams and contributing to the development of the nation. Teamwork and the willingness of local workers to do so are thus as much a commentary about their *active* participation in the development of the nation [in the spirit of gotong royong], as it is about their location as workers or organization members.

That this same willingness and symbolic currency associated between participation in quality activities and development are *precisely* [and paradoxically] what is *enabling* state planners, trainers and managers to actively harness and organize the labor and efforts of locals in the interest of institutional or economic aims is one of the great ironies and effects of power attached to the story I tell here. As such, while local managers' and officials' reliance on and appropriation of gotong royong to explain or translate modern notions and practices of teamwork and participation do offer locals a way of articulating potential alterities and subjectivities [to development, to organization, to workplace relations] in situations of inequality, this reliance and appropriation also has the effect of deflecting attention *away* from the potentially controlling and disciplinary features attached to these modern know how.

Concurrently, not only is local culture constantly mediating how the non-local is constituted, it is also actively appropriated and used by local officials in their encounters with both self and other, for example, local SMIs and foreign MNCs. Thus we find the concept of "Malaysia Inc." actively used to compel industry participation in the service of national aims.

To this end, it is not so much the "mediation of" or even "mediation by" culture that we are talking about here. Rather, it is the active "*re-presentation*" and re-constitution of culture that is on hand. Within this process, the very associations and meanings attached to cultural concepts or practices *shift* in relation to the specificities framing and organizing their very articulation. Thus, what are seen by managers and trainers as a way of translating the global to the local, may for *others elsewhere*—for example, government officials—be a way of reminding or "telling" yet more others (e.g., MNCs, local SMIs, citizenry at large), of the stakes and accountabilities they should be attentive to.

In this sense, culture itself, and the way that its presence is "made known" to us is constantly *shifting* in terms of what it signifies. It is neither fixed nor knowable until the instance of articulation, or unless situated within the social particularities constituting its very articulation. It is to this constantly evolving standpoint that I referred when I observed that stories told in the chapter are not intended to be taken as indicative of some endemic trait of either Malays or the nation alike. In contrast, knowing from this evolving standpoint makes possible the ability to see how knowledge itself [what is and can be voiced] is situated, or rather, negotiated into being. As in the case of my interactions with the Infotech manager and NPC trainer, it was not until a certain commonality or understanding [in location, in frame of reference and so on] had been negotiated that the multiplicity of meanings, practices, critiques and so forth "emerged." At times articulated, at times not, this dialogical knowing is significant, since what emerges often calls into question received notions about what "something is like" [that is, its taken for granted-ness], while bringing into focus alternative ways of understanding the local.

Chapter Seven

Counter-Representations and Negotiations for Voice

Why is Western or American knowledge global and ours local? This question was posed to me during one of many conversations I had with a professor and administrator from a local university when our discussion had turned to the knowledge-making enterprise and what this may entail in terms of knowing. I remained silent for what seemed a long time, although in real terms, it was probably no more than seconds. In any case, it really was not so much a question as it was a statement, made somewhat quizzically with a gentle reflexive-ness embedded in its very articulation.

And so, silent I remained. What indeed can I say? Or have to say? For I too, like this individual, often find myself asking this question. Not just in connection with my official designation as a researcher or academic, but also in connection with my other selves and my other lives. A silence, self-imposed and other-imposed marks my engagement to my writing and my profession. Why is our knowledge local and others' global?

In a sense, I suppose I could have given an answer. A list of I-could-tells that would stretch from a place faraway and a time long ago whose tentacles can yet be seen in the here and now, but I didn't. Besides, this person already knows most, if not all, of these I-could-tells anyway. So what's the point? Finally, with a little shake of the head, I gave my response, "I don't know really," I found myself saying.

What gives some the right to speak for others? Indeed, to speak over others? What is the basis for this right—a bigger voice, a louder voice, a proven track record? Record of what? Economic might, political right? Progress? Science? Efficiency? Where do we as diplomats of life and learning come from when we do something called "knowing" and "writing"? Or is it writing and knowing? Who do we speak as when we do so? Who do we speak for in doing so? And what are we hoping to accomplish when we know and write?

* * *

This chapter looks at the multiple sites of power and forms of engagement set into play as a result of changes being effected in Malaysia. Unlike the active appropriations, applications and mediations noted in preceding chapters, what is contested in the stories that follow are dominant constructions about locals' subjectivities and the meanings informing their action and intentions, produced here by discourses circulating in the nation as a result of changes being put into place.

As heart, these counter-constitutions are primarily a commentary against the implicit and explicit disciplining and neutralization of intent, of voice, of history engendered by the adoption and deployment of modern discourses, in particular, discourses governing productivity and efficiency, time use and effectiveness, quality and accountability. When looked at from this standpoint of knowledge and power, what we see being contested are the often narrow and reductive representations produced by these discourses whose overriding interests are economic and administrative in nature and aim. Before going on to consider the specific features of these contestations and counter-representations, I look first at the institutional procedures organizing local managers' constitutions of reality and of local workers from this standpoint of productivity and administrative efficiency.

* * *

When you have a prime minister telling you that your way of life and your cultural heritage are insufficient, civilized and genteel though they are, to see you through to the future, what is done to you? When this person and many others tell you that this civility and gentility is precisely why you're "done for" unless drastic changes are made to and in the self, what is done to you? When this life and this heritage become the basis of a never-ending series of psychological and sociological and economic prodding and biopsies, and you're told that you're a walking talking problem, what is done to you?

Concurrently, when you're told, no matter how hard you try to be a model nation in a modern universe, that you're still a problem, that you don't have what it takes, what is done to you? When you're told after trying very hard to play by the rules, which incidentally are not yours to begin with, that the rules have just been changed, for your benefit and well-being no less, what is done to you? And when you try breaking the rules as you see others doing, and you're told excuse me, but the exceptions don't really apply to you, what is done to you? And when you're told, on voicing your frustration with this game that is not yours, that you're evincing characteristics bordering on the paranoid and the psychotic, what is done to you?

When you're told, in acknowledging your fatigue with the demands made on you in the workplace, that it is simply because you are still

> *young, yet to mature or acquire the proficiency needed to manage either your self or your time well, and should therefore, and maybe, by the way, get someone you barely know to help you with your problems, what is done to you? When you cannot give voice to a silence that changes the safety and well being of workers into a conversation about routes and head count, what is done to you? When you have to ask why your knowledge is local and others' global, what is done to you?*

* * *

THE ADMINISTRATIVE ABSORPTION OF THE LOCAL INTO THE GLOBAL

In the example below, I attend in more detail to the procedures by which the local is not only rendered knowable, but is reconstituted into forms of knowing that silence the social particularities of the locality. In this instance, focus is on a company meeting held at Multitech to deal with the problem of labor shortage in Malaysia. It illustrates the often procedural recasting of social reality that occurs within everyday practices like planning events or meetings, and *how* this recasting is effected and mediated by the *already* organized forms of knowing and doing belonging to managers—professional, institutional—whose words and actions they are.

During the months preceding this meeting, increasing scrutiny over the abuse of foreign workers by local employment agencies—along with concerns over a variety of social problems linked to their rising presence in the nation—had induced state officials to clamp down on these agencies, many of whom were supply and administrative funnels for workers to the multinationals. Over this time, both Multitech and Infotech spent the majority of their time attempting to finds ways they could deal with this problem:

> **A:** [departments have] no reconciliation within their own respective units . . . the *forecasting* visibility here is for the next five months . . . there's a dilemma between production and production control . . . we're losing more than we can hire at this time . . .
>
> [*On a projector:* an overhead on "*Direct Headcount Recruitment Plan.*" A discussion about the potential scenarios attached to these follows. At some point, another overhead on "*Department Headcount and Requirements Status*" is placed on the projector. Listed here: information on "Headcount Requirement," "Headcount Available," "Need for Hire," and "Hired" for each business area.]
>
> **M1:** [starts to compare between the number of workers they can expect from one of their contractors, how many headcount they need to hire,

> projected turnover, and possible incentive packages to offer existing workers to get them to stay—this information he jots down on big sheets of paper on a stand]
>
> **M2:** Putting this on paper [M1's comment about notifying workers of the incentive packages] . . . there'll be a war—no, let's not put the actual dollar figure on paper . . . [discussion follows about how the contractor would feel about this. From here, it ranges over other agents they can turn to, the problems they could encounter hiring from different states, money that needs to be expedited to hire new people. Discussion returns to the proposal on what an incentive program may include.]
>
> [*Writings on white-board:* M1 jots down another figure, with information correlating "Months of Service" to financial rewards for both "Introducers" and "New Hires" to be paid over these times. Alternative scenarios of correlation and timing are discussed.]
>
> **M2:** . . . this may bring on another war—but it's a war we can afford to fight—still, it's less expensive than hiring [Indonesians?] . . . [a long pause] . . . which are more motivational? [reference, incentive packages] . . . right now, we're only talking about direct labor . . . basically we're using the "carrot-method" . . . maybe we can get them to stay longer . . . it is in their interest . . .

As represented here, the issue of labor shortage is primarily an administrative and coordination question about what managers can or should do to keep production going. Framed thus, the problem and solution confronting managers center around their ability to forecast and resolve the dilemma between needed and available labor.

To understand what is producing this specific constitution of the problem, attention to the categories of analysis used by managers in representing the problem such as "headcount requirement," "headcount available," "need for hire" and "hired" show how reliance on these categories effectively objectify the social reality of labor and labor shortages into standardized and numerical ways of knowing that allows managers to *factor* labor and labor shortage *in* as part of their planning activity. Specifically, the construction of workers via the concept of "headcount" systematically *recasts* them into numbers that can now be tabulated, cross-tabulated, forecasted and so on. This recasting is also conceptually mediated by concepts like motivation and references to the motivating power of money [*basically, we're using the carrot-method*]: mediations made possible by a presumption of workers as self-maximizing economic subjects who would or should be able to think rationally about this given the alternative [whatever this may be]. What is produced as a result is the possibility of formulating solutions in terms of a variety of incentive packages—indeed, the possibility of focusing on incentives as ways to confront labor shortages.

Seen in this way, the concepts and categories used by managers do not just reflect the reality of labor shortage or the act[ivity] of planning, they provide the conceptual and documentary *procedures needed for planning to take place.* Thus, planning events such as this meeting are key sites where social reality [and uncertainty] is reconstituted into forms that render problems amenable for treatment (Escobar, 1994b). In effect, what is produced through these conceptual and documentary applications is the systematic *reduction* of reality into standardized forms whose presence as standard knowledge provide to managers and organizations alike the conceptual *stability* needed to plan.

Thus, the overheads used during meetings and the categories inscribed on them do not just reflect this objectification of reality, they ensure that discussion is consistently organized around the concepts, categories and/or forms of knowing represented by them. In referring back to the example above, we see that conversation is consistently informed by what has been written-down: the overheads and their knowledge acting as conceptual and documentary forms mediating—and stabilizing—both what can be said and not said, seen and not seen.

In the process, what is rendered silent and out of sight are the social particularities organizing the phenomenon of labor shortage to begin with, whether in reference to government concerns with the abuse of foreign workers, tensions between local and non-local workers, or economic growth and labor shortage. Also notable for its absence is any discussion over *why* there might be such shortages or high turnover problems. In this sense, the concepts and procedures of objectification I bring to attention also make possible managers' ability to *abstract* from [local] accountability, any responsibility on the part of the institution, or selves, in potentially contributing to or producing such problems. In effect, the locality *disappears* in the face of managers' reliance on or use of these categories.

As noted by Escobar (1994b) and Smith (1987), the framing of sight engendered by the application of conceptual, documentary and textual forms and practices such as concepts and schemata are powerful institutional and professional procedures for abstracting the particularities of the local into a generalized knowing or known. What is produced is a generic knowing, standardized and so applicable to any and all contexts. However, this knowledge and power procedure is not just confined to the activities of managers or multinationals that I document here. It is part and parcel of our everyday practices as professional academics. Recall the conceptual bounding and mediation of Zambian workers' social reality via the concept of "motivation" noted in the earlier review of the cross-cultural management literature. To the degree that this is so, what I speak of is the reductive power of all professional forms of knowing and doing, whether managerial, academic and so on.

In sum, rather than reflecting some naturally-preexisting or naturally-occurring universality [organizing, planning, management], attendance to such interconnections between knowledge and power bring our attention to the *universalizing* effects of professional discourse and practice in establishing a context in point, and a form of analysis from wherein action and intention can be effected on an everyday basis, whether culled under the auspices of problem identification, administration, coordination, planning and the like. Put another way, the procedures and practices relied on to plan, to problem solve and so on provide the much-needed [conceptual and administrative] stability and standardization required for comparisons and evaluations, coordination and planning, to take place. Much like the application of QC tools by workers, the categories, schemata and rationality encoded in managerial and organizational practices are not just constantly involved in defining what people can see and how, they are simultaneously *producing modernity* at the level of the local.

* * *

> *Time and again upon my return from fieldwork in Malaysia, concerned and interested individuals would approach me and ask: so tell me about Malaysia, how was it? What did you find? Not realizing as they did so, that they were throwing me into a personal and representational conundrum. I scarcely know what to say, where to begin and how to say it. After all, there is so much one can say and know about the place. What are they interested in? What do they want to know? Where are they coming from? And where are they hoping I can take them?*
>
> *What, on the other hand, should I say about Malaysia? Where specifically should I start? What should I focus on? And where do I want to take them?*
>
> *Inevitably, I find myself doing what I have done time and again—rushing excitedly to convey a sense of the richness, the color and the complexity of a people, a history and a nation, all at once, in ten minutes no less, hoping as I do so that my enthusiasm alone will be enough to light the fire of their interest and appreciation for this place I used to call home. Or, crippled by the realization of the impossibility of ever conveying the richness, the color and the complexity of either people, history or nation, I mumble and stumble over some inane description and reply—ever aware, to quote Ruth Behar (1996)—of the injustice I am committing. Filled with the sinking realization, as she so eloquently goes on to describe, that when I packed my suitcases and left, I was not so much bringing a world back as I was actually leaving one behind—no less, indeed.*

This is not so much a confession as it is an acknowledgement about the limitations of knowledge; of knowing only from a distance, and having to communicate in and across distance (Behar, 1996)—even if this distance is mostly and finally a conceptual one. How does one go about writing and presenting the simultaneity that is a nation? Or the people inhabiting and inhabited by it? How to recreate in two-dimensional text, this simultaneity and multiplicity? To capture the life in a living history? Or, in Behar's words, to resurrect the ethnographic moment and communicate the distance?

Simultaneously a symbol of the politics of voice I was contemplating in the main text, it was on a personal and experiential level, a painful confrontation and struggle to avoid from being lost forever on a journey that started long before my entry into the profession; through a "tunnel" that I did not fully apprehend. The tunnel of an experience or venture that, once committed to, does not end until it ends (Behar, 1996). Where I too was convinced that this may well be a journey for which I will find no exit this time.

An egregious situation aggravated by a cretinous series of questions shadowing the writing process itself: how will I ever do it? Can I do it? How on earth did I ever think to even do it? What was I thinking of? Are people even interested anyway? Will they listen? Will they understand? When will it be done? When will the story be finished? How will I know? Indeed, will I even know? And where will I go from here? What other stories will I tell? Or not tell?

* * *

CONTESTATIONS AND COUNTER-REPRESENTATIONS

I turn now to the contestations and counter-representations articulated by workers in their bid to give voice to precisely those universalizing procedures of knowledge and power documented previously. For most of the stories recreated here, there is both an active contestation over, and appropriation of dominant claims framing people's tales to and in the service of voicing their own claims and experiences. However, it needs noting that the stories recreated are less a "direct" onslaught against the dominant than it is a commentary about the effects of power or struggles for voice endemic to or located in changes currently taking place in the nation. We find in these contestations a simultaneous and paradoxical commentary on power and agency: points noted variously by ethnographers of resistance (e.g., Comaroff, 1985; Ong, 1987; Taussig, 1980), and postcolonial historians (e.g., Chatterjee, 1993, 1988).

In considering the issues articulated by workers and managers alike, what we see is how concerns are often aimed at organizational constitutions of their time and labor, and the administrative requirements or procedures underwriting these constitutions and claims: namely, the discursive organization of everyday managerial or organizational activity that had produced such requirements and procedures to begin with. Let us look briefly at the specific example of time, participation and productivity. There is a great need within modern workplaces to know how time is used, what is being produced, who is participating and who is not. From the standpoint of this discursive regulation of sight and activity, the connection between time and participation exists because participation as constituted within the spaces of these companies is not just about standardizing or regulating workers' analysis and activity [noted in chapter five], it is also about the efficient use of time, which is productivity itself (e.g., Thompson, 1967; see also Clark, 1982; Hassard, 1999, 1990; Lakoff and Johnson, 1980; McKendrick, 1962; Nowotny, 1976; Taylor, 1911). Put another way, interest in participation and its connection to productivity exists because of the already commodified notion of time and labor that exists in the minds of managers and organizations alike.

As a disciplinary procedure, this focus on time and time-use regulates workplace and worker activity by providing to managers a means for tracking the actions and contributions of workers. They are barometers for assessing how people should be and how well they do what they are called on to do. As in the example of Multitech's TQM posters, the constitution of participation and of workers represented in them are not only highly objectified, this objectification is precisely what is allowing managers to track and *know* what or how workers are doing. Thus we see that participation is known in terms of "number of meetings held" and "attendance." This implicit constitution of time-use [numericized and economized] is also configured through multiple productivity and quality related indices used in these posters: for example, suggestions forwarded and implemented, actions and results, costs or defects reduced, production increases, and reductions in cycle times. As *barometers* of time well used, these representations rely on and produce a very specific constitution of what workers are—that is, productive resources—and how they should be—namely, participatory and attentive.

Concurrently, the possibility for contesting and counter-representing ruling notions about time, participation and the like arises from the existence of an already localized and alternative constitution of time, of self and of participation that provide to locals, the agential "space" by which to know and to articulate their critiques. This information about the representational aspects in regard to "time and the Other" (Fabian, 1983) is one that has been well noted by numerous scholars from the social and management sciences (e.g., Calás, 1992; Cavendish, 1982; Ditton, 1979; Fabian, 1983; Roy, 1960).

As I was to discover very quickly in the multinationals I was observing, time is a commodity constantly under contestation or in short supply. This is true for both workers and managers. For managers, it is an essential component against which they are constantly running. Phrases and comments by managers like "there's not enough lead-time," "the response is too slow," "time is very short," "while mucking around with this, time is already lost," and "this should have been done already" construct for members and audience alike a sense about the scarcity of time: a feeling that things should have been done yesterday, regardless of input and/or effort. For workers, the battle is one over how to get all their tasks done, often in the face of contradictory expectations and increasing pressures resulting from specific organizational policies such as time optimization.

Take the following construction of his staff members forwarded by the human resources manager from Multitech in response to increasing discontent over both the shortage of staff in the department [kept lean and mean by management in the name of efficiency], and problems resulting from this shortage, such as overwork and stress. Over the weeks preceding these conversations, members of the department had literally been scrambling to do [and undo in certain cases] the multiple tasks that they are responsible for, often putting in eighteen-hour days for long stretches at a time:

> . . . Y and Z need to be able to do some time management, they're both very stressed out right now . . . I told Y, "talk to Vanessa—she's a third party, maybe she can help you to prioritize" . . . I've talked to Y, but not Z . . . both are fairly new—they're not used to how things go here . . . time management . . . they're putting lots of stress on themselves . . . they think I'm a monster . . . but stress, I am used to it.

His assessment situated already within the company's official aims [time optimization, efficiency], the manager, while acknowledging the stress amongst his staff, forwards a construction of them as individuals lacking the ability and capacity to manage their time well. As in our earlier example of definitions forwarded by officials about young Malays and SMIs, these constructions entail an active definition of both Y and Z as young and inexperienced, and by implication, unused to the normal requirements and discipline of the modern workplace.

In doing so, the manager reconstitutes their experiences as *individualized* problems [*time management, they're putting lots of stress on themselves*], rather than symptomatic of the company's lean and mean policy.

That is, as problems traceable to the personal inability of these workers to either cope with or meet the demands required of them in their capacity as workers and organizational members. Ironically, he himself implicitly acknowledges the potential reality of their claims with his reference to the stress that he himself often experiences. What is neutralized in the process are the very concerns of these workers, making it possible for the manager to conceptually and discursively *abstract* [via the focus on time management] from local accountability, any responsibility on the part of the institution or the managerial self.

In the following, Y provides her "take" on this problem during a subsequent conversation with me about the "lean and mean" policy of the company's. She pinpoints attention on the multiple tasks she is now responsible for, and the lack of recognition by management over this contribution of workers:

> . . . [I've been] manually entering data from it [reference a survey conducted to track lockers], someone else should do this [list of additional activities she has been doing] . . . long hours alright if the recognition is there . . . as sometimes the efforts are all for nothing . . . talked to [the HR manager] about this . . . am taking care of tags, [list of additional add-ons] . . . am I a worker or executive?

To Y, the issue is not having enough time to do the things she needs to do, and the extra amount of responsibilities she has to carry—something that she does not mind if the recognition is there. The contestation here is one over the *value* of her time and contribution. The implicit critique: the lack of understanding from seemingly unresponsive policies and managers. It is a frustration that leads her to re-appropriate the very constitution of her time and boundary of responsibility by bringing attention back to what she is *supposed to be* doing in an official capacity [*am I a worker or executive?*]. In the process, she calls into question her manager's constitution of her since the issue for her is less one of time management as it is the organization's hiring policy and management's insensitivity to workers' plight.

These counter-constitutions are subsequently reinforced by one of her colleagues, a senior member of the department who has been with the company since its inception. Like Y, she calls attention to the added load borne by workers, highlighting the call on her time that this entails; a scarcity that means she often cannot get to what she *is* supposed to be responsible for:

> . . . We need more people, additional people, both A and B are new, right now I am constantly training . . . I cannot even change their status to clerks, only recorder . . . even though all have access to confidential stuff

> . . . they [top management] just don't know, they don't know what it's like . . . people are all leaving, so I keep training [pauses, lowers her voice, switching between English and Hakka as she goes on] . . . *aiyah* Vanessa, you don't know . . . my reputation is so bad so bad, everyone looks at me and they talk you know, "what is X doing? How come this is not done?" I keep telling [her manager], "we have to have someone on stable, just when they're trained, they all leave . . . " [pause, shakes her head] . . . sometimes, it is just crazy*lah* . . . we have only cosmetic measures, there is no quality effort . . .

In addition to the value of their time and contribution, what is contested in the examples above are notions about self and subjectivity. Through the active re-appropriation of official designations regarding their locations and work responsibilities, these workers simultaneously pinpoint attention to sites and forms of power regulating their actions and self, while actively *re*-presenting their own contributive capacity and productivity, for instance, their unstinting efforts to take on or juggle additional demands.

Rather than the unproductive or inexperienced workers that management presents, they call into question the aptitude of managerial and organizational efficacy. In the above, X does so by actively re-appropriating the company's own symbol of efficiency and reflecting it back in the form of a judgment that the company and its management are composed only of "cosmetic measures," not "quality effort." The conceptual and documentary forms relied on by managers and organizations to know and assess workers [time-use, productivity, quality] are actively appropriated in the service of contesting dominant constructions.

The conceptual and administrative neutralization of experience framing workers' contestations is especially evident in the following account of how managers from Multitech are confronting operators' concerns over the safety of the company's buses, and the recklessness of the drivers ferrying them back and forth to work every day. Over the weeks preceding the company discussion documented below, there were numerous reports in the local media about problems with the safety of factory buses from other companies: a situation aggravated by the report of a case where an operator was raped by a driver on her way home from work in the early hours of the morning. At Multitech, while managers had been aware of this concern for some time, the responses tabled in the face of this increasing concern such as the dissemination of a questionnaire, and spot checks by security and administrative personnel did not seem to have much effect on the situation. The following is an excerpt from a company meeting on this issue:

> **M1:** A survey questionnaire among operators has been circulated . . . they wanted better buses and air-conditioning . . . apparently, the buses are very rickety [and fast! Someone adds amidst a few chuckles and groans] . . . [the chief of security] has got some of the questionnaires . . . maybe we should get someone to sit in once a week . . .
>
> **M2:** We have done it before . . . there are some problems along certain routes . . . [the chief of security] is planning to get on all routes . . . also, the operators wanted more stops [some comment here on safety stops which I didn't quite get] . . .
>
> **M1:** These are all good plans . . . sometimes we commit but don't follow through . . . most [operators] complain that the buses are too small . . . most are concerned for their lives . . . they keep talking about their *keselamatan,* their *keselamatan* ah [safety in Malay; a few chuckles at the repetition of the word, which in the Malay language, can be a way of injecting humor into situations, behavior and so on, a form of ridicule.] . . . [pause as chuckles and gaggle of conversation subsides] . . . there is no way that there can be an overload, there are very many routes, we should not be exceeding capacity . . .

While safety is an acknowledged concern, what is accomplished over the course of discussion is the systematic reconstitution of this problem into a primary concern with logistics and coordination. This procedural recasting is produced via M1's reliance on singularly objectified concepts like "overload," "number of routes" and "capacity" to represent workers' concerns with their safety. In the process, the potential social, political and economic issues connected to the problem are reframed as a technocratic question.

Moreover, it is a reconstitution that forwards to those present at the meeting, a construction of workers as *not knowing* what they're talking about. This dismissal of concerns [*there is no way there can be an overload, there are many routes, we should not be exceeding capacity*] that operators had brought forth in connection to the safety of buses—their physical conditions, problems along certain routes, number of stops, size of buses—is a specific instance of how workers and their voices are summarily silenced by knowledge, what Radhakrishnan (1994) calls the negation of Others through knowledge. Specifically, as the Other [the known] of managers' knowledge, workers' concerns are subordinated and juxtaposed against the implied rationality and/or superiority of managers' [the knowers] administrative know how and experience: a hierarchical evaluation made possible and reinforced by the "poke" about operators' concerns over their *keselamatan.*

Juxtaposed against the constructions of subjects and issue in the above is the following re-presentation of this problem, made to me the next day, by one of the assistants from the same department:

> . . . We did cut off one route but the drivers are reckless, and the buses . . . there are umbrellas inside, the windows are stuck . . . I keep telling [them they] "have to do something" but these [transportation companies] are old relationships with the company and they don't want to break it off . . . but I am responsible for the safety of the workers here . . . the buses don't run on time . . . their engines break down . . . you know, I yelled at them to ensure timeliness . . . there is interest in the long run to do so [doing something], maintaining the stability of labor and timeliness . . . they have to do something higher up [pause] but there's quite a lot of pressure too . . . we are so stretched here at XYZ department, I am surprised that much more has not struck yet . . . we need additional helpers . . . so this is the pressure and nature of HQ policies in [Western country] [pause] . . . I am mad because nothing gets done . . . we have no contract thus far with the bus company, only a verbal agreement, so anything can happen . . . they [administration, organization] talk about *semangat sivik* [Malay for "civic responsibility"] about TQM . . . where is their *semangat sivik* now?

Encoded in her words is an acknowledgement over the existence and immediacy of this safety problem. Not only does she give voice to the legitimacy of workers' claims, she renders visible certain information that had been absent in the meeting noted previously. Specifically, the recklessness of the drivers, the poor conditions of the buses, the company's reluctance to terminate relations with its transportation company, pressure from headquarters and shortages resulting from this.

Also articulated here is a certain frustration with the company for *not* having taken any action, and an explicit voicelessness over the issue [*I keep telling [them they] have to do something*]. While an active attempt is made by her to explain the reasons for such non-action [*but there's quite a lot of pressure too*], this continued non-response is to her nothing short of civic irresponsibility. Rather than simply just administrative and so on in nature, her concern is the safety of workers and the potential problems that can occur in the face of continued non-action by the company. Put another way, the problem *for her* is the implied existence of a subject [higher ups, management, company] who seems *incapable* of either taking action or appreciating the dire seriousness of the problem they are confronted with.

In forwarding this active counter-constitution, she brings *attention back* to the official claims and policies of the company [quality management, teamwork, responsibility and accountability], and *reflects it back* in the form of an acerbic observation regarding *their* inability to keep to their *own* policies and claims [*they talk about semangat sivik, about TQM, where is their semangat sivik now?*]. Thus, loaded-into her very reference to and juxtaposition of civic responsibility and TQM is an explicit reminder about the ethos of teamwork and participation that so often attend managers' constitution and deployment of their [quality] management activities. The critique forged here is one over the [seemingly hypocritical] *lack of* social responsibility that managers and company alike are showing through their reluctance to take action of a substantial nature.

What is significant to note from the standpoint of how the local [culture, meanings] mediates understanding, is her specific reference to civic responsibility and TQM *to* this inaction. Given the association between TQM and community-oriented notions of teamwork and collaboration noted in the previous chapter, this allusion to safety and managerial action *as* an issue *about* TQM and civic responsibility serves as a reminder that all are *already framed in her mind in terms of* the community-focused, jointly accountable and collaborative notion of gotong royong. Her critique is, in this sense, very much forged from *within* the localized and cultured understanding of what TQM and teamwork are all about to begin with.

* * *

This story has, in a manner of speaking, been a long time in the making. It started when Malaysia became a place to go to instead of a return home. Launched into the disquieting role of diaspora as an adolescent, I made a promise to myself that one day I would return. While I did not know in exact terms under what circumstances, or in what guise this return would take, return I would. It was, I suppose, a way of negotiating the act of leaving itself. Although, I did ask myself at the time whether the promise had been made in earnest, or simply in a vain attempt to justify my leave-taking. The intention and politics fueling this claim and this promise, an attempt to answer a question I had asked myself then [why am I here?], sustaining the activities and choices I made as an adult.

And so return I did. In the self-described and institutionally inscribed role of "bridge" between diaspora and home, east and west, north and south, researcher and subject.

Only to find in the instance of yet another leave-taking that the story I have to tell can never really be the one I want to tell. In the confrontation between experience and data, data and representation, representation and text, text and profession, I lost the Malaysia I had gone to find, and found instead a subject and a subject-location.

From the purview of knowing as opposed to living, "Malaysia" can only ever exist as an object of interest or an entry-point for inquiry. As a living entity and a living history, it can only ever "be"—by virtue of how we know and why we know—in relation to the gaze and the aim of those looking for and those looking at it, its location and presence as a living entity and history, peripheral to its location and presence as an object and subject of investigation. It cannot, and does not in this sense,

> *exist outside of this gaze and this intent belonging to the observer: bound forever in its location as "the observed."*
>
> *As a person inhabiting this divide between a "living-Malaysia" and a "subject-Malaysia," this fissure is more chasm than gap, simultaneously and paradoxically separated and linked by a universe of concepts and convention.*
>
> *As a researcher, knowledge of the conceptual and instrumental foundation making-up the field of experience and action, while easy enough to understand from a theoretical point of view, becomes on a personal front, a stricture structuring sight, where the pursuit of a goal mediates against divergence from the prescribed script on how-to and what-to and why-so, even as the living experience of the person would indicate that there may be and is a connection which can be made between the professional and the personal, the official and the unofficial. Yet the language of the professional does not seem to allow for this.*

* * *

The explicit frustration over the lack of voice, and struggle to negotiate voice that I document above is one that is heard not just within the confines of multinationals. It is also a sentiment articulated by numerous officials and administrators in their constitution of changes taking place in Malaysia [corporatization, deregulation, quality management]. While most see the state's concern with and need to undertake these changes, what some question is *the process* by which such changes are being put into place.

The following is an example of such a questioning, forwarded here by a senior administrator from one of the local universities during my conversation with him about the initiative to corporatize the educational sector:

> . . . So okay, it has become a necessity now to think global. And we have to go along with it . . . and come up with our own objectives . . . going back to the dynamics of the change process . . . it comes from the vice-chancellor, and the university has its own interests in it . . . the staff and the academics . . . the staff and the union, they say "we need to be consulted in the process." What is there to be consulted? Everything is coming from the Ministry! There is nothing to say . . . students are also coming into the act—being sensitive to the rise in tuition fees mainly . . . I tell them "quality should go up" [b]ut none are talking about the quality of education [pause] . . . we have employers coming and saying that the quality of our graduates are not as good as foreign graduates . . .

Seemingly caught to some extent between the multiplicity of interests calling on him in his location as a senior administrator, he not only calls attention to the existence of these multiple claims, but in doing so, brings our attention to the process by which the intended change is being effected. What is questioned here is the very process of change [*what is there to be consulted?—everything is coming from the Ministry!*]. Specifically, he highlights the lack of voice and participation between the state, which is promulgating such changes, and the rest of the community (e.g., staff, academics, students, union), including the administrators of these universities.

This critique over the lack of engagement and consultation is reinforced by his reference to the concept of quality, a key precept under which such changes have been constituted. Framed by his preceding comment about the lack of consultation, this reference back to official constructions *for* change amounts to an explicit reminder to state officials [and others alike], that changes currently being put into place do *not* quite reflect the premises under which they had been forwarded [*but none are talking about the quality of education*]. To this end, he actively re-appropriates the market-oriented language of officials and reflects it back in terms of the reminder that quality is what matters. A contention reinforced by his comment that *employers themselves* [the customers and market for the university's products] are voicing the claims to quality that he is highlighting in his indictment about the process of change.

Thus, contestations noted here and elsewhere are not just a simple negotiation over time, action or intention. Embedded in the stories are contestations over the subjectivity and agency of individuals (e.g., workers, managers, officials) whose actions and lives are being actively reconstituted by discourses and changes introduced in the nation. Constructed against official policies governing their use or value of time, participation or productive capacity, these contestations are often effected under the auspices of local culture, and through the active appropriation of dominant claims. As noted by Escobar (1994a), the potential for "counter-labeling" is always attached to the labeling and naming activities of institutions. This is particularly so given the administrative and economic reductions that are such an endemic feature of modern institutions of rule: reductions that can only ever render the objects of its knowledge (or gaze) knowable in narrow and simple terms.

Workers and officials quoted challenge official constructions about their subjectivities and contributions while also calling attention to the potentially neutralizing effects (e.g., the silencing of contribution, meaning, voice) attached to the forms of knowing and doing informing these official constructions to begin with. Similar to the spirit attacks that Ong (1987) documents in her ethnography of female factory workers in Malaysia, references to gotong royong and the attending focus on civic responsibility, process, voice and collaboration that we find in the words of workers and officials above are a commentary on the *assault* on locals' notion of community and social responsibility, indeed, on engagements between self and community, resulting from the introduction of economized, objectified and organization-centered forms of participation and collaboration in the nation.

In this regard, the focus on community and collaboration represented by gotong royong is a way of understanding the operations of power and powerlessness resulting from changes being introduced in the nation. Gotong royong *itself* is in this sense, a specific attempt to articulate an alternative development and subjectivity; one founded on mutual engagement between self and community, participation and voice.

* * *

And so it was that I found myself marginalizing that which I claimed to hold dear.

The mirror is not only cracked; it is unforgiving in its ability to refract the fragments of a self, so that what can only ever be seen is a fragmented reality, shattered by the institutions and locations the self, all selves, inhabit. For in the very instance of articulation, a world had already been shut out and lost.

And it is only at the point of return, when Malaysia, so to speak, has become yet again a place to go to, where the distinction between experience and real time, observation and historical time, is greatest that this fracture between experience and data, living and knowing, is most visible and real.

The ethnographic moment becomes a fleeting experience of reality; the reality literally an ethnographic "moment." Abstracted from this moment, a nation and a way of life lose its reality and meaning. And attempts to recreate it seemed more an illusion than real.

Which should I privilege? What is sacrificed when I choose one over the other? On what basis must I and can I then rest my story?

Empathy and a critical awareness at one end of a scale and conventions of reason and science the other end pulling at this silence and this voice; a balance looking for equality and birth.

* * *

Thus, while my earlier interest was triggered by a need to consider how know how transferred is changing the terrain of the everyday [what is put into place, what people are asked to do], my interest here is in what locals' encounters with this know how can tell us about its [know howow's] "status" and location as organizers of social life and action: in short, their disciplinary features and effects (e.g., Foucault, 1979, 1980; Ong, 1987). In this sense, focus on the multiple forms, content and sites of contestation provide to us an opportunity to better understand what precisely is problematic about these knowledge and practices and *how so*.

In the example below, the description of Infotech's participatory activities and their links to the company's aims provided by the human resource manager brings our attention to the narrow, organization-centered personas [subjects] considered desirable by the company, and the pressures [indeed, *constitutional possibility*] resulting from a commodified notion of time and its use:

> M: . . . And then you have the "personalized" ah, the "personalization" of this shared vision [shaking his head a little as he said this] . . .
>
> V: How do you have this personalization? [somewhat puzzled and curious as I asked this]
>
> M: [he chuckles, leans forward and answered, demonstrating with his hands what he was describing and explaining to me: curving them around an imaginary space representing the vision and placing them in front of his heart as he responds] . . . somehow, they put this "vision" in front of them and make it their own!
>
> V: [I grinned, thinking about the modern personas of the "organization man" or "woman"—not at him, but (maybe?) with him.]
>
> M: . . . But we are not there yet [shaking his head] . . . we only have "shared" vision . . .
>
> V: Oh by the sound of it, it doesn't seem like Infotech is doing too badly . . .
>
> M: You think so? [still leaning forward as he said this] . . . we started this . . . oh . . . maybe three years ago, we are not there yet . . .
>
> V: [I pursed my lips and frowned a little] Mmm . . . that's pretty fast no? Considering how large the organization is . . . it might be different if you are a small company with only ten or fifteen employees . . .
>
> M: [he grins, shrugs, leaning quickly backwards and forwards again] yeah? Speed of Execution! [one of Infotech's objectives] . . . not enough speed . . . three years . . .
>
> V: [I grinned back] . . . yeah . . . I guess we'll see huh?
>
> M:[he smiles, shaking his head]

Implicit above is a commentary about organizational procedures of objectification and subjectification: specifically, the reliance on participation as an effective way for cultivating subjects who are literally personifications of the company and its aims, and the presumed capacity on the part of managers [as official representatives] to identify and to know when such an end goal has been achieved. In this case, there is an assumed notion of "being" that is *procedurally* identifiable, fixed and measurable [via time-use, turnaround capacity, speed].

Equally significant is how the manager conveys this information. Framed by his seeming mockery of the company's intention, and his doubtful responses, the association of the company's objective to his escalating "negativity" can be read as a veritable questioning about both company aim and expectations. As noted by scholars writing on power and resistance (e.g., Chatterjee 1988, 1985; Ong, 1987), an endemic commentary on how power is experienced—and resisted—can often be found in subaltern appropriations of dominant symbols as an avenue of resistance; where resistance itself is also a commentary on power. In the example above, the manager is actively appropriating the company's official objective in the service of conveying *the opposite* to what was officially intended: a critique reinforced by his repeated shakes of the head, even in the midst of communicating what are seemingly positive responses.

As I noted to myself when I was writing this fieldnote, there was something in the manager's tone that suggested skepticism, whether of "personalizing" itself, or the time it was supposedly taking. At the time, I was more inclined to believe that he was just that bit cynical about this possibility, and that he was in fact "parroting" what might have been said to him by others about the time it was taking to accomplish it. To this end, his specific focus on speed and personalization cue attention to the regulation of *his* time, his actions and his self by these organizational stipulations.

Reclaiming the Self by Reconstituting the Other

It is not just in reference to organizational or policy actions that contestations I present above are forged against. It is also a feature of locals' encounters with specific concepts being deployed or used to represent and constitute them, whether as a culture, a people or a nation. In the two examples I document below, there is a concerted attempt by both individuals speaking to reconstitute the location of the self as subjects and recipients of Western discourses: dominant constructions that relegate them to an implied position of inferiority, whether as an object of censure or transfer. Within both acts of reconstitution is a concerted re-appropriation of those concepts and values making up and making possible this hierarchical constitution of their selves to begin with.

In the first example, I return to the words of the NPC quality control trainer I referred to in the previous chapter. While my intent there was to illustrate how local notions of participation mediates the deployment and adoption of quality practices, my intent here is to highlight the trainer's re-appropriation of the concept of "flexibility" in the interest of counter-representing constructions of the self which were less than complimentary [*they think it's like selling medicine in the night market*]; not to mention the implicit presumption that an *authentic* QCC form exists. Forged against these boundaries of constitution and demarcation, the trainer's description of local QCC practice not only valorizes the self [*rather than black and white, we have color*], his very description of local QCCs actively constructs an association between them and the discourse's central claims to flexibility and teamwork [*for Malaysia, we have to be a little more flexible, circles may be as large as 15, numbers is not an issue; for us, we have to be a little more flexible, some may be a year, may be more, this is the Malaysian way*].

Under this counter-representation, it is in *locals'* ability to adapt, adjust or circumvent the *in*flexible conventions and precepts of official QCC practice that a commentary on what flexibility is can be found. The trainer is not only actively re-presenting a construction of the self as shaman-like medicine sellers in night markets, or non-authentic QCC practitioners, he also *reclaims* the location of the discourse as a knowledge and practice of flexible management and collaboration, albeit mediated *this time* by the local forms of QCC practices.

Concomitantly, what is also actively being re-constituted is *an Other* who is distinctly *non*flexible and *non*-collaborative in their privileging of efficiency over participation, numbers over adaptability, length of presentation over the involvement of members. The trainer brings our attention to these multiple forms of "absence" by resorting to words like "strict[ness]," "restrict[ion]," "precision," "straight to the point" and "black and white" to describe and counter-represent Others and their inflexibility. In effect, he reconstitutes them as the very opposite to what they see themselves as being or representing. In doing so, he changes the very meaning of flexibility itself, and how or what it is. Flexibility in this case, is the local capacity to adapt or appropriate an otherwise non-flexible or non-local form.

Indeed, according to the director of one of the skills development centers, the focus on "management by collaboration" in quality discourses is "very Asian." As in the above, we see the concerted *reclaiming* of knowledge *and* location here. This reclaiming is reinforced by his subsequent assertion that such a collaboration, and the implied advantages attached to it, is something which the West is "only just realizing"; an assertion that he repeats later in our conversation:

> . . . Asians . . . like the Japanese, they never tick off their employees in front of others . . . [it's] very important to save face . . . *Americans are buying back some of these ideas* . . . companies like Infotech, they take on the frontline on leadership and they call it "watching each other's self-esteem" . . .

Rather than knowledge transfers *from* the West *to* the Rest, what is forwarded is the very opposite claim. Namely, current interests in quality management and the adoption of these ostensibly Western or industrialized practices are seen instead as nothing more than a "*naming*" procedure [to appropriate Adelle Mueller's (1987b) concept] since this focus on community, collaboration and teamwork have always been an indelible component of local (Japanese, Asian, Malaysian) practices [*Asians already doing this without realizing it*]. If anything, these know how are actually signifiers of a West learning and appropriating *from* the Rest [*Americans are buying back some of these ideas*]. In the process, he actively reconstitutes the location of this self from object or recipient of knowledge to promulgator or owner of knowledge, albeit heretofore an "un-named" knowledge.

In sum, the stories recreated here bring our attention to the active *re*-presentations (Radhakrishnan, 1996) of Self and Other informing locals' constitutions of their experiences and encounters with knowledge being disseminated in Malaysia. A process often mediated by their locality and location, these contestations involve the active appropriation of concepts and values of ruling discourses as a way of giving voice to or criticizing the particular forms of power currently organizing their actions and intention. In this sense, contestations do not just pinpoint attention to what is problematic about managerial practices forming the core of locals' critiques; they focus on how and why these practices are problematic.

In the process, they remind us about what is *not* immediately apparent in the specific vicinity of events, encounters and procedures they are speaking to, and the silencing or silence that attends what is officially going on or being accomplished. As in the case of the worker from Multitech highlighting management's lack of action regarding worker safety, her counter-constitution and critique remind us that there are certain factors bounding or organizing action and speech whose *very absence* in everyday discussions are what makes possible the actual administering or planning activities of managers: a commentary on how power is sometimes equally derived from what is said as it is from what is *not said,* or kept-out of everyday purview and actions.

* * *

Silenced thus, the researcher I entered a paralyzing struggle to articulate an experience through a form and a lens that cannot seem to allow for such an acknowledgement, let alone engagement and dialogue; organized all the while by the institutional conventions of academe and a deep-seated socialization into the norms of the profession. I kept wondering to myself why it was that I couldn't write-in as opposed to simply writing-down. A world of difference seeming to reside between the "in" and "down" of writing.

With two committee members left holding the baggage of a profession and an institution within which they too are simultaneously caught and reside—mustering from goodness knows where enough grace and mindfulness to tell the storyteller I that she still has a voice and a message to convey when I needed it the most.

Truth with a capital "T" presupposes the non-necessity for dialogue by virtue of its status as, and claim to Truth itself. What's distinctive about Truth is the absence of experiences other than its own: an impermeability and imperviousness to conversation. How can processes and accountability be written-in within a context such as this? How can "experience" speak—however this may be understood—if the very negotiation for voice is one forged, and must necessarily be forged, no matter what its intent, on the very premises and foundation of theory and method itself?

* * *

THE PROBLEMATIC OF ARTICULATING THE LOCAL IN A GLOBAL

The two examples to follow highlight the power attached to ways discourses constitute and organize people, even as they signify the existence of a knowing subject who is all too aware of how they are being organized. Both circulate around an implicit or explicit confrontation with the implicit forms of knowing and doing encoded in management knowledge and the construction of this knowledge. The first example involves a conversation with a business unit manager from Infotech. Triggered by a joking comment of his, conversation turned to what his education background involved. It is a turn that led eventually to the following commentary on the nature and capacity of management knowledge to "know":

V: How did you find the MBA program, did you like it?

M: [tilts his head, thinks a little, purses his lips] . . . mhmm . . . yeah . . . yes [pause], but some of the things are very simple ah . . .

V: [I look at him inquiringly]

M: [he leans forward in his seat, using his hands to emphasize his point as he answers] . . . some of the behavioral analysis is very simple isn't it? The things they teach you . . . you take a problem [gesticulates with his hands] they keep taking . . . they take some variables away to simplify the analysis . . .

V: [I look at him intently, nodding now and then]

M: [pauses and shakes his head] It's a simplification isn't it? . . .

V: Yeah, yeah . . . it can be very controlled in its analysis . . .

M: Exactly [he nods] . . .

Constructed against an amorphous "they" [management scholars, educators], his critique focuses on the deductive and reductive features of modern management analysis and education: modernist features signified here via the concept of "simplication." It is a term he uses repeatedly in his attempt to communicate the reductive knowledge taught to him in business school. Also implicit in his description is a question about the capacity of management knowledge to capture or represent phenomenon itself. Here, he actively appropriates the realist ontology and functionalist epistemology informing the field's claims to knowledge and uses it to comment on the knowledge's *[in]ability* to "properly" represent reality: hence, his repeated emphasis on simplification.

Framed by this critique, what follows is a discussion about what or how else management [analysis, knowledge, understanding] may be; triggered here by my question about what *he* thinks it takes "to manage":

M: I think you have to have some rule of thumbs [pause] . . . if you're dealing with situations . . .

V: [I nod, looking at him]

M: It's something more [nods, pauses, holds his hands up] u-uhmm . . . psy-sychological [he says slowly, frowns, pauses] . . . if you apply it in real life [reference to whatever he was referring to earlier], it cannot be done . . . things are not as clear-cut as that [speech pattern and tone more "emphatic" as he continues, saying the last quite quickly and firmly]

V: Yes . . . yes . . . things are more complex than that . . .

M: [he nods, sitting back a little more in his chair] To manage it, you have to manage it personally . . . sometimes it is difficult ah . . . you only know when you are face to face with someone . . . then you can tell . . . how the other person is . . . whether they are honest or . . . how they are . . .

V: [I nod] How do you tell?

M: [he nods] Like I said, you have to have rule of thumbs ah . . . based on your past experience . . . s-so . . . understanding human behavior . . . you have to understand people from a psychological standpoint and . . . it's from past experience [pauses and grins] like all the contingencies . . . and contingencies they talk about all the time!

Of note here is the difficulty he had as he attempts to describe and convey what and how he manages. Marked by a series of hesitations, pauses and frowns, he literally could be seen grappling with how to describe what he means in his attempt to articulate what and how he knows. Emerging finally, albeit hesitantly, with the concept of "psychology," he nonetheless continues to seem unsatisfied with this choice of concept to represent his meaning, instanced here by yet another pause and more frowns.

Part of the difficulty he seemed to have with the concept may come from the recognition that within the management and behavioral sciences, the fields of psychology and social psychology (e.g., Argyris, 1957; Katz and Kahn, 1966) have been crucial in producing and up-holding the very know how that he is himself critiquing here. In this sense, the use of the concept to represent what *he* means is potentially problematic given this historicized bounding, especially so in the face of his preceding critique.

For my part, his search for words to articulate his meaning was paralleled by my own attempt to read where he is coming from, what he means and the connection between what he is now saying and his earlier critique: a potentially paradoxical exercise given the bounding of his description by concepts and assumptions such as "rules of thumb" and "psychology" that on first encounter would seem to contradict and subvert the very foundation of his critique, although admittedly, this was a potential contradiction and paradox which he himself seems to be quite aware of.

It was not until the sequence following immediately after his elaboration that this tension was resolved somewhat:

V: [I start to chuckle] No, yes . . . yes . . . [pause] Do you think this is very typical of . . . here? MX [another manager] and I were talking about some of these things the other day . . . if there are differences in how we—in Asia, whatever—how we approach or deal with things . . .

M: [he nods a little] Yes, it is very much from your past experience . . . and the rules of thumb [smiling as he adds] you have to "*tai-yan*" ah . . .

V: [I nodded and smiled] Yeah, sometimes I find it very difficult to explain to my friends in the West . . . that management [rather, knowing?] is not just about . . . uhmm . . . variables or . . . or guidelines [pause]

> . . . that there may be something more you know . . . s-some . . . oh I don't know . . . some intuition [said in a rush as I struggled to articulate what I was trying to describe; cringing somewhat even as I said this] . . . like ah . . . you know, the "*qam-kok*" that you have! [waving my hands a little and turning them around in little circles as I was trying to find my words.]

Conveyed in Cantonese, tai-yan, literally translates into "watch the people," whether in reference to looking or observing. Translated on a more contextual level, it can also mean "depending on the person," or "be sensitive to different persons." In other words, a way of acknowledging and representing the *non*-universality of being, indeed, the presence of another [human] being, and a reminder about the context surrounding behavior, interactions, encounters, intentions and so on. Hence, his earlier contention that knowing itself *cannot* really be known ahead of time, that it can only be known "personally," "face-to-face" and "contingency-like." From this standpoint of attentiveness to difference, references to the "personal" is not so much a commentary on individuality as it is one over the *particularity* of what and how knowing is accomplished.

These meanings and constitutions are doubly significant in light of his earlier critique about the simplicity and difficulty of reducing human behavior to the level of models and formulas. *Within* this situated knowing (e.g., Haraway, 1988; Harding, 1998; Rich, 1984), his references to rules of thumb and past behavior becomes less some teleological statement about fixed rules than it is an acknowledgement about the engagement of the observer-knower self in establishing a context in point, an interaction, observation, judgment, action and so on. From the perspective of the researcher I, this knowledge about *his* "particularity" [and location] changes in meaning the very concepts appropriated and used by him (psychology, rules, contingencies). Although it needs reminding that such concepts, given *their* own histories and located-ness, are also potentially powerful ways of circumscribing what can be and cannot be said. Hence, the difficulty he had in describing what and how he approaches management, and mine in locating the nuances and meanings attached to his attempts at description.

Indeed, I myself was subjected to a similar quandary. Indexed by the struggle I had in finding the words [whether English or Chinese] to describe what I mean, I found myself simultaneously torn between the need to articulate the how of my approach, and circumscribed by the vicissitudes of management and behavior as they have been discursively constituted by the field. Additionally, my discomfort came less from any lack of knowledge of the language itself than it was in having to use it *in association with* a concept and practice I seldom think of in Chinese (i.e., management). Thus, I was equally torn by the need to try and describe how I know, and to convey my agreement with him in his reference to the concept of *tai-yan,* settling finally on the concept of *qam-kok* [a Chinese concept, also spoken in Cantonese, which can literally mean intuition or feeling, even instinct; under certain circumstances, it can also have a more negative connotation, as in foreboding]. In the context of what he was saying, it was *my* way of acknowledging the difficulty of saying itself.

This implicit "surround" of the field and its conventions and knowledge [his job, my arena of study] that is organizing our exchange is explicitly noted by him immediately after in the form of a cautionary reminder:

> M: [he grins as I was saying this; turning it into a smile and nodding at the same time] [chuckles] Yeah, but you can't write this down! ah . . . [continues to chuckle while shaking his head, bending forward a little in his chair]
>
> V: [I laugh in return, shaking my head] No . . . if you write it down, they're gonna think you're crazy or something!
>
> M: [he turns his shaking into a nod, still grinning] Yes . . . but in terms of technical training ah, they're very good . . . Western training . . .

Articulated amidst some amusement, his words nonetheless remind us that our exchange and resorting to community-bound concepts not only contravenes the mainstream in its constitution of reality and knowing, it does so via a critique of certain practices and norms [of knowledge, of science, of legitimacy] belonging to the respective institutions we reside in whose very articulation or documentation—while giving voice to the knowledge that organizes our worlds—would entail a certain confrontation with the foundation on which these concepts were premised. To this end, the sequence of exchanges above not only calls attention to the forms of knowledge and power constituting the subject under consideration *and* our discussion of the subject, they tell a story about the potential difficulties of actually articulating what the self "knows": a self- and other-imposed silence often marking engagements between this self and this knowledge.

Looking for Islam in "Islamic Management"

As juxtaposition to the above, this next example illustrates the paradoxes attached to attempts at giving voice to a self by relying on languages not one's own. On the one hand, it speaks to the "absorptive" power of modern knowledge, while on the other, it highlights the difficulty of speaking outside of this language that is constantly framing and circumscribing what is seen, what can be said and how.

In various discussions I had with some colleagues in Malaysia on the proliferation of material and interest pertaining to Islamic and Confucian-based management approaches that are circulating in the nation, the following comment was made to me by one such colleague in his assessment of these writings, and what they are attempting to articulate as a body of knowledge during an interview I conducted with him:

> But the thing is . . . the Japanese also do this [i.e., manage according to their own principles], and so too the Western manager. Also, the consequences remains the same . . . so what's so different about it? It's management . . . it then just becomes another version. . . . where's the religion then? But even if it's different, it is on a religious basis, to be practiced only by those with the faith—how can you transfer that? So how universal can it be?

Starting with an implicit recognition about the located-ness of management practices, his assessment quickly gets caught-up in what Kate Manzo (1991) calls the "modernist trap."

Central to the seemingly paradoxical "circularity" that he finds himself in is an acceptance and reliance on singularly modernist constitutions and criteria for evaluating knowledge itself. We find as an instance, that his assessment of this knowledge is already bounded by norms of universalism, and the implicit privileging of secularity over spirituality [a feature associated with universal modernism's belief in science and rationality], applicability and transferability over local capacity or meaning [*even if it's different, it is on a religious basis, to be practiced only by those with the faith, how can you transfer that?—so how universal can it be?*], and end-result over process [*the consequences remains the same, so what's so different about it? It's management*].

Framed within such criteria, knowledge is both legitimate and valuable *only* in the context of what it can achieve, and how amenable it is to application and adaptation *elsewhere.* There is a certain functionality in judgment that is already written into his assessment over its [implied] viability. Hence, his concern and reference to "transfer[ability]" as the *barometer* for evaluating the knowledge's claims and value.

Although it needs noting that he himself, whether intentionally or unintentionally, is also partially pinpointing attention to a modernizing procedure through which such a "silence" is produced. Encoded in the specific sequence that I refer to here [*the consequences remains the same, so what's so different about it? It's management, it then just becomes another version, where's the religion then?*], is a commentary about the power of modernity's economized constitution of reality to organize sight and judgment. Framed by his references to "consequences" and "management," his subsequent assessment—bounded by this implicit economism and its extended focus on results, performance and so on—*make possible* the very abstraction of the local (e.g., religion as a basis for managing or organizing) into a "global" knowing (management) and evaluation of knowledge (universality, transferability). What is silenced in the process is the existence and locality of the knowledge form he is assessing.

The example highlights the power of modern concepts and procedures to abstract the specificity of the local into a general such that the local is systematically de-linked from its social and historical particularities (Smith, 1987). In particular, it calls attention to the difficulty of seeing or voicing *anything other than* what is specified by and within the ruling discourse once the categories of analysis, schemata, values and assumptions upholding it are accepted and used as the starting point of analysis.

Thus, we find that the "Islam" attached to "Islamic management" is consistently defined, judged and organized *around* and *in relation to* the ruling concept of management: forever in the shadow, unable to emerge or speak in the presence of this concept which is constantly privileged over it. It is a point reminiscent of Rushdie's (1991) notation about commonwealth literature, and the consistent relegation of writers not from the center to a position of "subalternity" whose location can only ever be defined in relation to this center—the epistemological equivalent, in some ways, to Mahathir Mohamad's constitution of the nation's economic defense strategy.

SUMMARY

Stories recreated here and elsewhere are not just about the active reconstitution of self by locals. They are equally the tales of a people and a nation caught between multiple words and multiple worlds; torn between a need to acquire and so defend the Self against an increasingly global "language" on the one hand, and an equally pressing need to negotiate for voice on the other, belonging simultaneously to both worlds and neither at the same time. Yet, the very possibility for such a story as I tell here is also testament to the existence of a locality and a local agency that, *in spite of* this concerted bounding of voice, constitution of subjectivity, and organization of action and intention, will insist on being heard. In this sense, the active appropriations and counter-representations recounted in the stories are themselves a signifier of a local subject only all too aware of how their lives and actions are being organized, even if such recognition is often articulated *within* the boundaries of a dominant claim organizing both lives and action.

It is partially in reference to these dynamics that I made the observation in my Introduction that Malaysia is, metaphorically speaking, a world "at war." Where the researcher and knowledge, state planner and plans, manager and organization, worker and workplace, are all lenses by which this war is fought: where right is the power of voice, and power the right to voice. In itself, possibly a prognosis that is more than apt and less than kind since inherent in the stories to be told is a multiplicity of representations, struggles and meanings to be gleaned. And so it is [and was] that the story that is or can be told is as much an index of the researcher, the state planner, the manager, the worker and their respective lenses, as it is about a Malaysia [waiting] "there" to be documented.

* * *

And so it is that I often find myself wanting to tell innocent and curious inquirers alike: go, go see for yourself. I am not the expert and the eyes by which and through which you should see. Not because I have neither base nor claim to expertise, but because expertise itself can only be silenced through its own non-existence. And so, go and see. But before you go, be sure to take with you all your lenses: intellect and learning; memory and history. And please, don't forget the subject locations you bring with you.

Or better yet, stay still and see. There is little need for an "elsewhere" when the object of this seeing is the self.

But the story, you see, it doesn't end there.

By acknowledging and opening up for consideration this feature of knowing and knowledge making, by speaking directly about the living experience of being absolutely involved as an agent of knowledge and power, I open up to scrutiny the very story I tell. To quote and appropriate [or is it misquote and misappropriate?] Behar's words, on a scale with "tragedy" on one end, and "farce" the other, the choice seems less an option than it is no-option. Torn on the one hand by the tragedy of having to [re]-fashion for the self a story that this self can only ever understand in partial terms, and pulled on the other by the farce of having to stand in front of an audience, [re]-consolidating the self's uncertainty and dependency on subjects long gone and far away into a position of authority yet again as the self now stands in front of her peers, and forwards what is but a partial glimpse of truth into a full scale and unmitigated Truth, with a capital "T."

As you read these lines and these words, made up of my thoughts and my feelings, what do you see? Do you see in them a history and the traces of a life both real and imagined? Do you see in these words and thoughts the stories I have told throughout this text: a nation and its struggles, a people and their voices? Malaysia with her gnarled old trees lining boulevards leading people home, the schoolgirls with bulging knapsacks and old men on becas—will you see in them the story that is I/eye? And in me the story that is them?

This world of mine, my family, my encounters with geese, my conversations with friends, my re-acquaintance with nation, relearning of history, they are here in the words I write and the stories I tell.

Will you see in these lines and these words the boundaries of knowledge and the limitations of knowing—the limitations of knowledge and the possibilities of knowing? Or will you see in both the non-existence of knowledge and knowing? In stories, the farcity of partial claims to truth as opposed to the reality of Truth? Will you think these thoughts and these feelings an exercise in futility and incomprehensibility? Where are you? Where are you from and where are you heading to? How are you in this story that I tell? Where are you in this story that I tell?

In the words of Thich Nhat Hanh (1998), object and subject are one, two sides of the same conceptual coin.

What is your silence? What is the story you are not willing to tell even as a story you tell (Behar, 1996)? Tell me. I want to know. Because every time you speak, I or others elsewhere and a world are being organized by you.

As I organize yours.

Chapter Eight

Discussion and Conclusion

In September 1998, Mahathir Mohamad, Malaysia's Prime Minister at the time announced to the world that the nation would, in its bid to deal with the economic damage wrought by the Asian Economic Crisis (1997–1998), henceforth tighten and re-regulate the state's policy governing financial markets and the movement of international capital in and out of its borders. According to him, rather than resolving the crisis, restructuring policies forwarded by international development agencies had:

> exposed and bankrupted not only bad companies but it had rendered good companies non-viable. . . . [putting] banks and other financial institutions in danger . . . [and] forc[ing] countries to borrow heavily from international agencies. . . . [compelling] proud independent countries . . . [to] surrender their economic and political freedom (Mohamad, 1997: 9–10).

Responses to this announcement came fast from the international community, including representatives from development agencies, and analysts of one kind or another from industry and government. Their reactions and observations ranged from the very personal to the political and ideological. Despite this range in their responses, there was seldom any serious attempt to look into the concerns or meanings informing action and intention at the local level. If anything, observations of the local tended to be somewhat dismissive. For example, an article in *The Economist* (1999) represented locals who were bent on supporting Mahathir Mohamad as non-principled entities motivated *solely* by economic security and self-interest. This representation is at once ironic and paradoxical given that the article was contesting precisely the very existence of such modern market citizenship in Malaysia.

Equally significant was the generalized and implicit sense of surprise, indeed, of shock, that met Mahathir Mohamad's announcement. These were understandable reactions, given that those economies hit by the crisis—Malaysia included—were only a few years ago taken by most in the West as barometers and symbols of the New World Era. These were nations heralded by agencies like the World Bank (1993a) for their economic, social and political initiative in "freeing" their respective economies from the yoke of non-market and regulated forms of development: instances as it were, of the efficacy of neoliberal policies to development. Mahathir Mohamad's announcement, coming as it did on the heels of such optimism and projections, offered instead what seemed to be an indelible challenge to the viability of neoliberalism *as* a development policy.

Rather than simply non-reflexivity regarding the historically located "surround" organizing relations between the West and the Rest, these reactions and observations signify an equally problematic issue governing relations between the two. More specifically, this account of unmet expectations, contesting interpretations and counter-constructions triggered by the economic crisis is symptomatic of the limitations attached to received approaches to knowledge, and the representational practices informing the production of these knowledge. What is called into question by this recent example is the implicit belief that we have here in the West about the *knowability* of Others, and the assumption that they can be *predicted through knowledge.*

In this particular instance, the knowability of nations like Malaysia was one already limited and bounded by knowers' [development agencies, management theorists] reliance on a coterie of narrow and singularly economized categories to identify and locate these nations (e.g., World Bank, 1993a). Moreover, this assumption of knowability and representational practice of knowing through self-interested and self-referential categories were compounded by the assumption that *once known* [or located thus], these representational Truths can be taken, *once and for all,* as the final designator and predictor of—and for—whole cultures, societies and nations.

It is to these practices and assumptions informing our knowledge that I referred when I observed that encounters between the Self and Other are first and foremost a question about our approach to knowledge and understanding about others. It has been a central feature of the story I tell in this text that the knowledge we now have about Others, and the practices and relations they represent are key arenas in need of revisiting and revising in our ongoing attempts to configure and reconfigure relations between a knowing Self and those Others who are the objects of our interest. It was with this intention of reconstituting a different form of knowing and doing, albeit via a focus on *current* forms of knowing and doing, that this project had been formulated. What I have sought to do with—and in—the telling of this story on knowledge transfers was to explore and provide for potential alternatives in how we can know and what we can do. What I mean by alternative is "knowledge" able and capable of letting be, and of recognizing the shifting simultaneities that make up phenomena and our knowledge of such phenomena, while remaining accountable to the presence of the knower self in his/her engagement in this perhaps different process of knowledge acquisition and knowledge construction.

As an example, the a priori assumption of universality that is connected to our knowledge and its claims shut out from consideration alternative voices and understandings about the phenomenon in question. It is a constitution that leaves little room to account for the possibility of *engagement* between these subjects and this knowledge since others are seen primarily as devoid of agency and volition. That is, as passive recipients rather than active agents of knowledge and history. As a result of such assumptions and beliefs, we find in most received scholarship on knowledge transfers a conception of transfers as a literal and linear process originating from the West to the non-West (e.g., Gupta and Govindarajan, 1991; Osborn and Hagedoorn, 1997; Park and Ungson, 1997). In the instance of the economic crisis, this lack of conceptual and methodological "space" for incorporating and representing alternatives in Others' understanding and voice has rendered it difficult for us to understand the localization of global practices and knowledge.

In the following segment, I look in greater detail at how the story of transfers as told in preceding chapters contravenes and questions this possibility of knowing and predicting Others through conventional knowledge. This is followed by a closer look at the implications and issues that this story presents to both the managerial and knowledge production enterprises in the terrain of the organization and international or cross-cultural management. Specific attention is focused on how the framework and methodologies informing my telling of the story have sought to address and redress these issues. In the final segment, I re-situate the project in terms of current efforts in the field of organization studies aimed at searching for and articulating alternatives in scholarship, in research and in representation [textual and non-textual] that are able and capable of configuring knowledge equally attentive to self and other, and the engagements underwriting relations between the two.

THE STORY OF TRANSFERS

Contrary to the story of a universal and unified process presented in received scholarship, the institutional ethnography of knowledge transfers I undertook in Malaysia highlights that it is a non-linear and non-literal process marked by a mutually constitutive series of engagements between the local and the global/western. More specifically, it is a process informed by a series of appropriations, contestations and translations that are mediated at different points by culture, by memory, by history and by social particularities surrounding the phenomenon that are specific to the nation. Currently, there is a growing awareness and an emerging body of scholarship by numerous actor network researchers both in and external to the field of management (e.g., Gherardi and Nicolini, 2000; Law, 1999, 1997a, 1997b; special issue of *Organization,* 2000) pinpointing attention to precisely such translations and reinterpretations that "mark" the "social ordering" process (Law, 1999: 860). Or as noted by Gherardi and Nicolini (2000: 344) in their study on the circulation of safety knowledge, this "circulation and constitution of organizational knowledge [i]s the process and product of the translation made by an action net comprising individuals and communities, organizations and institutions, and their intermediaries . . . which connects the local to the global, the practice to the theory and vice versa."

In the case of Malaysia, local managers and trainers involved in disseminating quality control knowledge call-on the rural Malay concept of gotong royong to convey to locals the meaning of teamwork and participation as it is commonly understood in modern workplaces in the West. Noted previously, this is part of the mediatory and bridging activity these individuals are required to perform as a result of the cultural crossroads they inhabit [*we have to be "two'*]. From the standpoint of translation, what is significant is how this process is one *already* bounded by the cultured terrain and space that translators and local concepts are situated in. It is a "bounding" that changes the very meaning and practice of knowledge transferred. Thus, like the translations noted by actor-network theorists, the actions, meanings and intentions of subjects I note here are constantly being constituted by the network and engagements of the actor[s] to and within this network (Law, 1997a).

Accordingly, we find that the emphasis on community welfare and privileging of process over end results that is embedded in the concept of gotong royong not only provided a means for translation to local managers and trainers, it transformed the actual meaning and practice of teamwork. This reconfiguration in meaning *shifted* attention away from the organization-centered and economized constitution normally associated with QCCs towards a more process- and community-oriented interpretation. As a result, what we find important to local workers are their membership in teams, and the opportunities to be fully contributing members of teams, communities and/or nation; rather than simply a focus on efficiency, productivity, performance and the like.

This example of "having to be two" and the focus on translation as an activity that comes already located by culture and pre-existing meanings counters the assumption that knowledge is easily, or indeed, "transferable." This is so since the meanings and forms of concepts and practices being transferred are forever shifted and changed *in* the process of transfer [rather, translation]. There is no fixedness to either concepts or meanings since these can only ever be known in terms of the specific locality organizing the translation or interpretation process and act (e.g., Haraway, 1988; Rich, 1984). This focus on the ever-shifting nature of meaning and the difficulties or consequences attached to attempts by knowers to "fix signification" has been noted by, and forms one of the key insights of poststructuralist scholarship (Calás and Smircich, 1999).

Concurrently, active *appropriations* of knowledge also contributed to the existence of concepts and practices whose meanings and forms are forever changing. For example, state officials in Malaysia were actively involved in appropriating features of neoliberal prescriptions and discourse in the interest of community and national welfare. In this instance, while policies governing more open trade and deregulation were adopted and human resource development activities implemented, their adoption and implementation were organized less in the interest of free-markets than they were forged in the service of *defending* the nation against free-market policies in general, and the economic domination of an industrialized elite in particular.

Unlike the assertive and predictive representation offered by the World Bank (1993a), the story reconfigured here about Malaysia's adoption of market policies contravenes, however paradoxically, this tale of neoliberal efficacy and viability. On the contrary, it pinpoints attention to the *already counter-constitutive* act marking the nation's adoption of these policies. To this end, the issue of primary concern—and significance—is less the act or fact of adoption as it is the implicit and explicit commentary on power and voice, inclusion and exclusion, which such a counter-constitution as I recount here is addressing.

To further complicate the picture, local officials' appropriation of these policies is, like workers' and managers' engagements with the concept and practice of teamwork, already mediated by the social purview of gotong royong. Used actively to call for greater social responsibility by business to the welfare of community and nation, officials' attempt to articulate an alternative development is already located in this privileging of a more community-oriented approach. As a consequence, the very meanings and existence of business-government collaborations in Malaysia are recast from being a neoliberal symbol of deregulation and privatization to a signifier [and practice] for cultivating a more socially responsive and one might add, defensive, alternative. Thus, as in the case of translations, this mediation of the global by the local, and the already counter-constitutive act marking transfers of knowledge counter the possibility of assuming *a priori* the identifiability and/or knowability of Others.

Rather than fixed and predictable, what we see in this story so far is the existence of a *multiplicity* of meanings associated with concepts and practices forming the objects of transfer. Thus we find that what is an official multinational strategy for "motivating all employees in all countries" [TQM and teamwork] is, in this case, a signifier for a form of participation that is either distinctly localized or counter to the intents it was assumed to represent. In effect, what we see are instances of how local culture, and the situated interests of local officials are actively contributing to and engaged in changing or reinventing the meanings and practices of concepts like teamwork, collaboration and neoliberalism as we would know them here in the West.

But this infusion and engagement[s] of the local *in* the global is not the only factor shifting the meaning of concepts and practices transferred. We find for instance, that the mediation of *the global* in the local is also changing the very meaning and form of what gotong royong is now taken to be in the nation. Specifically, the appropriation of this previously rural Malay concept by local managers and officials in the service of conveying the production and work-oriented focus of modern organization practices like participation means that the concept is now firmly *linked* to the modernist auspices of both economy and workplace. The previously social and non-economic form of engagement and exchange between people and community represented by the concept is being systematically transformed and commodified by this shift into the terrain of official production and economic organization. It is moreover, a transformation effected and reinforced here by the specific reconstitution of locals into modern economic subjects (labor, workers, human resources), and the more generalized economization of social life brought on by the introduction of neoliberal development and management knowledge in the nation.

In a like manner, the appropriation of neoliberal discourses by local officials is equally marked by this mutuality—indeed, *simultaneity*—in constitution and transformation. The appropriation of these discourses, in particular the discourse of human resource development, is changing how officials and locals are coming to see and identify themselves. People are no longer seen or identified solely in terms of membership in their respective communities [Malay, Chinese, Indian], and whatever this may imply in terms of class, social status, religious affiliation and so on. They are now to be seen as human resources and workers: that is, factors of production whose identity and value are now to be judged not in relation to their community affiliations, but in accordance to their value as productive resources capable of contributing to the transformation of self and nation from developing to developed entity.

What the above signifies is the importance of attending to knowledge and power as producers, mediators and compellers of change. In this instance, the power of modern discourses to constitute how locals see themselves and the reality around them (e.g., Foucault, 1980, 1979; see also Escobar, 1994a, 1994b; Ong, 1987) is leading in the case of Malaysia, to yet another level and form of transformation—*contemporaneous to* and *existing in tandem with*—the changes produced by local culture and the politics of location detailed previously. Not only are concepts and practices non-fixed in meaning and form, so too are *identities*.

As a result of such changes and transformations, it is increasingly difficult to "fix" both meaning and identity as ways to know and locate Others. This is particularly so in instances where notions of identity are confined to compartmentalized and static designations on culture, on ethnicity and so on, which, as I observed in my review of the cross-cultural management literature, are singularly the researcher's to begin with (e.g., Hofstede, 1987, 1985). Put another way, the often self-referential categories relied on to know (e.g., whether people are collectivist or individualistic) are seldom attentive to the possibility that people may not see themselves as such, or that people may change in how they see themselves. This is so since assumptions of knowability and identifiability are already *preceding* the researcher's entry into the field and/or act of seeing. What we find being effected instead is a transformation in how locals themselves are seeing the self and others around them as they incorporate into their [pre]-existing notions of self [whatever or however this may be], increasingly modernized and technocratic forms for apprehending and knowing this self.

Thus to valorize identity solely in accordance with potentially essential or static categories in a bid to locate Others is to ignore the changes that are occurring as a result of this interpenetration of the global into the local and vice versa (e.g., Radhahrishnan, 1996). As I show in my documentation of the mediatory effects of culture, nations like Malaysia are increasingly constituted by discourses from the West, whose addition to or engagements with, pre-existing forms of identification are leading to the production of *multilocal* and *multifocal* subjects. I argue here that the voices of these increasingly hybridized subjects can only be rendered understandable and knowable in terms of where they are coming from, who they are speaking as, what they are talking about and so forth (e.g., Haraway, 1988; Rich, 1984). Or as in the example of the manager from Infotech whose subject-location (Malaysian nationalist, Malay, manager, muslim) was constantly shifting depending on what he was attending to (global capital, poverty reduction, social responsibility), my ability to locate and render understandable his speech and actions were directly connected to my ability to *situate* the context and social particularities surrounding his speech and actions.

Concomitantly, this story about the simultaneous processes of constitution and transformation [in meanings, subjectivities, identities] highlights an equally problematic tendency to locate and hold-still Others through knowledge. I refer to instances where researchers from the West, in bids to be more locally sensitive in their search for "native understandings," end up reifying once more those "local knowledge." For example, a researcher coming from this position may infer, on the basis of my writings and representation of this story, that gotong royong may now be taken as the definitive locator for Malaysia, and that Malaysians are fundamentally collaborative and consensual in nature and character.

This is no less problematic than instances where researchers are *in*sensitive to difference and diversity, for as I indicate above, gotong royong—while a central component and feature of social life—is already in the process of being transformed. *It is no longer what it was or might have been.* Moreover, as noted previously in my observations about cultural changes taking place in the nation, the prominence of the concept is as much by design as it is a commentary on some pre-existing meaning(s). In this case, it was a concept and practice actively resuscitated and appropriated by local managers and officials in *their* bid to communicate and translate modern notions of teamwork and collaboration to locals.

Thus, while we [researchers] may have become more informed and more sensitive to the nuances of a locality, it is equally important for us to remember that this same locality may well be [or is] already in the midst of being transformed; whether by dint of its own exigencies, its encounters with knowledge from elsewhere, or indeed, by its encounters with the researcher self. This potential to reify Others or privilege Otherness is a point well noted by Jayati Lal (1996) in her critique of the tendency amongst certain scholars to "fix" and "fetish-ize" Others in their [well-intentioned] bid to either reconstitute a more culturally-sensitive knowledge, or to turn the traditional binary of knowledge/power between Self and Other on its head. According to her:

> . . . the presumed epistemology of the native insider. . . . [is] essentialist and reduce either the native or Third World woman to an assumed homogeneous entity. . . . Furthermore . . . [these] constructions have the unintended effect of reinforcing the very distinctions that they are supposed to erase (Lal 1996: 198).

Put another way, to locate or know Others, whether via the reliance on generic [universal] or reified categories is to avoid taking accountability for the engagement of the researcher or institutional self in this process of constitution *and* reconstitution of "Otherness."

Accordingly, what is called for in this book is not a renewed endeavor to uncover some "true essence" belonging to others. The challenge to western knowledge and knowledge construction can no longer be resolved, for instance, by interpretive anthropology's approaches to understanding "the native's point of view" (e.g., Geertz, 1983) since—as I argue above—this native is no longer an "exotic other" waiting for us to discover. Rather, the challenge pertains to our ability to account for local representations and voices in the context of *our* knowledge and knowledge construction practices, and the equal need for us to consider the Other as active subject-agents, engaged as much as everyone else, in the construction of our contemporary world. The absence of such connective-reflexivity (e.g., Radhakrishnan, 1996, 1994) means that not only are differences and Otherness constantly in danger of being naturalized, it also means that the discursive and material bases of inequality, knowledge and constitution, power and voice, informing relations between this Self and Others are silenced yet again.

In sum, the story of transfers I tell in this book shows how the meanings and forms being transferred are constantly being shifted and changed, producing a multiplicity in meanings, in practices and in identities on the one hand; and constituted by a series of shifting simultaneities on the other, for instance, between local and global, knowledge and identities, meanings and practice, knowledge and power. Whether informed by local culture, by history, by location and indeed, by knowledge itself, this story questions in particular, the possibility of relying on fixed, compartmentalized and static categories in order to know. In doing so, it implicitly and explicitly challenges both the assumption that Others are knowable, and that "transferability" itself is possible. In the following segment, I consider the implications of this story, and look in greater detail at how it is connected to certain issues informing received understandings and practices of knowledge construction and production. Specifically, I focus on how the approach and methodologies informing my project have sought to address and redress these issues, and the implications of doing so.

CHALLENGES OF THE STORY

By attending to multiplicities and transformations, the foregoing brings to the fore issues governing research process and the challenges to representation and voice raised by the existence of this multiplicity and subjects' multivocality. Specifically, it challenges the notion that managing or researching in the international or cross-cultural arena is reducible to a question about the ability to know or predict Others. As the story shows, the contextual multiplicities and transformations disallow for such a possibility. How does one manage in the midst of such differences and shifts in meaning or practice? How to research phenomena that are neither fixed nor fully knowable, but constantly evolving in character and form, composed simultaneously of processes of constitution and contestation? Indeed, how does one go about understanding or documenting processes that are dynamic, involving people and institutions that are neither passive nor empty of meanings? What must the manager and/or researcher be attentive to in contexts such as these?

At the most basic and obvious level, the implication of this story is that we [manager or researcher] cannot take for granted that which is seemingly familiar, since this familiarity may have attached to it vastly different meanings and histories. For a manager or researcher operating under the assumption that knowledge from one locale is transferable to another, conversations and encounters with Others over concepts and practices like teamwork and participation can only be had in the midst of potentially grave misunderstandings; for example, the assumption that they are non-cooperative, inefficient or non-committed.

The story in this book tells instead about the mediatory power of the local, and the agency of subject-Others who come already filled with meanings, yet bounded all the while by an interlocking web of relations and practices that situates them in very specific ways to us. Within this story, rather than knowability or predictability, what can be seen are a myriad of engagements between the Self and the Other, and how these engagements are mediated and informed by very particular practices and relations of knowledge and power between knowers and known (e.g., Chatterjee, 1987; Escobar, 1994a, 1994b, 1988; Foucault, 1980, 1979; Mueller, 1987a; Ong, 1987; Prakash, 1990; Radhakrishnan, 1996, 1994). Knowledge is, in this sense, not so much predictability about Others, as it is the practices of constructing for ourselves, the *knowability* of our objects and subjects of interest.

In short, the issue is about how Others are *made known* or *knowable* by and to the Self. It is about how we know and how we construct knowledge. As such, the problem of knowability that I speak of is centrally a question about representation and voice. Recent deconstructions of the development sciences (e.g., Crush, 1995; Escobar, 1994a, 1994b, 1988; Mueller, 1987a; Sachs, 1992) and postcolonial historiographies (e.g., Chakrabarty, 1992; Chatterjee, 1987; Harding, 1998; Prakash, 1990; Radhakrishnan, 1996, 1994) provide an entry-point for considering these questions of representation and voice. In doing so, they call our attention to the interconnections between knowledge and power, and how these interconnections come culturally located [organized around values regarding rationality and efficiency, informed by assumptions about reality and objectivity], and embedded in practices [reliance on categories and schemata, binary oppositions] and relations of power [knowers and known].

To the degree that the researcher self and researched other is an endemic component of this relation, the issue is equally one about recognizing in the self, an *active agent* of knowledge and power. From the representations forged about Others to changes in identities being effected as a result of the introduction of modern discourses into the local terrain to the social or economic activities of multinationals and institutions like the World Bank, what we see is the presence of a Self who is actively engaged in socially organizing and re-organizing the subjects and terrains we study, or in which we practice. Within this nexus of institutional and institutionalized activities and engagements, what becomes significant is the need to be attentive to how the Self is involved in constituting and reconstituting Others through knowledge.

Here is where Smith's (1987) notion of thinking in terms of *relations*, and postcolonial historiographers' approach of thinking *relationally* (e.g., Chatterjee, 1987; Radhakrishnan, 1996, 1994) are critical entry-points for engaging with this myriad of engagements between Self and Other. The focus on relations reminds us about the interconnected nature of the worlds we live in and fosters awareness about how and where this self may be in relation to the world around us. Through its focus on knowledge and power as the social organizer constituting relations between self and other, it cultivates an awareness about the self's *location* as a knower and knowledge constructor. In doing so, the methodology brings the researcher back to those social particularities organizing how the self is as an agent and subject of knowledge, and what this constitution may entail in terms of what we do and our engagements with others.

It is this understanding about knowledge and how the self is that renders particularly important the re-constructive aim of this book to actively "write-in" the presence of the researcher and knower Self. He/she is a mediator between reality and our apprehension of it, and a social site by which knowledge is fashioned and constructed into being. Put another way, what this attendance to the presence and engagement of the knower accomplishes is to *situate* knowledge in terms of "who is speaking" and "where this speaking is coming from" (e.g., Calás and Smircich, 1999; Escobar, 1994b; Foucault, 1980, 1979; Haraway, 1988; Radhakrishnan, 1996, 1994; Rich, 1984; Smith, 1987).

Whether immediate or institutional, such attentiveness affords not only the opportunity to be accountable to social conditions giving rise to knowledge; it also provides a more situated knowledge about *Others* (Radhakrishnan, 1996; see also Calás and Smircich, 1999; Escobar, 1994a, 1994b; Smith, 1987). Thus we see as an example, that the actions of local officials in Malaysia are organized not in a vacuum, but come located instead within a social and historical terrain, constituted simultaneously by local and global factors like the history of the nation, investment policies of multinationals and so on. By attending to these social particularities organizing others, it is possible to transition *away* from practices of knowledge construction that serve only to reinforce potentially essentialized notions about them (e.g., Chatterjee, 1987; Lal, 1996; Radhakrishnan, 1996). In the case of transfers in Malaysia, rather than naturalizing its so-called underdevelopment, attention was paid instead to the variety of factors and relations that are producing this condition and our "knowledge" of it.

As noted previously, the World Bank's (1993a) presumption of knowability and reliance on self-referential categories to locate nations like Malaysia rendered its knowledge incapable of either accounting for the already counter-constitutive act informing the nation's adoption of neoliberal policy, or anticipating the prime minister's announcement about re-regulating policies governing financial markets and the movement of international capital in and out of its borders. I argue here that it is precisely this reliance on self-referential and self-interested categories that are not allowing us to be attentive to the already counter-constitutive act, or those issues and concerns *informing* this counter-constitution.

Such disregard for potential differences in meanings [whether counter-constitutive or not], the situated-ness of Others, and the accountability of the Self silences not only Others' voices, it also shuts out from consideration potentially crucial concerns that are promoting those very shifts in meanings and simultaneities to begin with. In the case of Malaysia, their pre-specified knowing led to the inability of knower(s) to attend to crucial issues of representation and voice, inclusion and exclusion, which were articulated by local officials in their various appropriations and reinterpretations of neoliberal development policy. To local officials, it was precisely this lack of [self]-accountability by an industrialized West, and the narrowness of representations forged about Malaysia, that were called into question by them.

Thus, rather than passive recipients of knowledge, Others are active agents and subjects of knowledge. They are constantly engaged in reclaiming their subjectivities and reinterpreting either the claims of dominant Others to know, or their location as objective and neutral bystanders in possession of some superior knowledge (e.g., Chatterjee, 1993; Ong, 1987). Following from the example above, officials in Malaysia (e.g., Mohamad, 1993a, 1993b) were actively involved in counter-representing the oft-assumed and oft-accepted notions about free markets, competition and free trade as representations of a fair and level playing field. They bring our attention instead to the often-unequal terrain and form that is the global free market and the webs of power and control that are the industrialization or development policies of the West.

In sum, attending to engagements between the Self and Other, including awareness over how the Self may be involved in constituting the terrains and subjects we study is important for understanding the changes in meaning and simultaneous processes of constitution and contestation highlighted in this project. Specifically, I show that there are ways—theoretical, methodological and textual—to take accountability over, and to reconsider these engagements that our current approaches and understandings of knowledge are either inattentive to or unable to let be.

This attentiveness to how we know is especially important given the situated and multi-focal nature of our objects and subjects of interest. Rather than the existence of a homogenized entity, what my story shows is the already historically and culturally located and increasingly hybridized terrain belonging to Others, whether in reference to meanings, voice or identities. What is called into question by attending to these differences is the possibility of any easy attempt to codify or categorize actions, cultures and nations that are constantly changing and/or coming into being. As an example, stories told in preceding chapters highlight the impossibility of reducing Malaysia or Malaysians to either homogenized wholes easily represented and equal in their social and historical formation, or to the constitution of nation and polity as a *tabula rasa* de-linked from and unaffected by the actions, decisions and histories of the West. In this regard, attending to the politics of location (Rich, 1984) is *not* just a requirement of being attentive to the Self, it is *equally* a requirement of being attentive to Others and the plethora of differences in location constituting them.

REPRESENTATIONAL ACCOUNTABILITY, PROCESS AND MULTIVOCALITY

What the framework and methodologies informing my project have provided is precisely this possibility of reconstituting a form of knowing and doing that is capable of being attentive to the Self and Other, including the engagements and relations informing encounters between the two. Specifically, institutional ethnography's (Smith, 1987) focus on relations and intersections, knowledge and power, and social particularities has been important for configuring a knowledge that is both accountable and capable of situating differences and change. Through its focus on the problematic of the everyday, and subsequent constitution of phenomenon *not* as phenomenon, but as an object and subject for inquiry, institutional ethnography is akin to poststructuralists' focus on the non-fixedness of meanings (e.g., Calás and Smircich, 1999). Neither accepts as prima facie what is knowable or apparent, and both question knowability itself as a condition of knowledge.

In this regard, the methodology's focus on Self and Other in terms of their specific locations *as* knowers and known—as opposed to senders or recipients of knowledge for example—implicitly and explicitly situates phenomenon like knowledge transfers and our knowledge of it, *in* practices and relations of knowledge and power. Accordingly, the approach and methodology I use in this book come from a very specific understanding of knowledge. It sees knowledge not just as empty or objective carriers and vehicles for conveying information such as tools, forms, schemas and practices. Rather, knowledge is composed of practices and relations that are cultured and institutionalized into our everyday activities as members of professions, as researchers and so on.

Within this frame of understanding, knowledge construction itself is a cultural activity (e.g., Calás and Smircich, 1999); embedded in knowledge and power and informed by values and assumptions organizing it as a way of knowing and seeing. In this way, the approach questions the implied neutrality of all knowledge-making enterprises and activities, and looks at knowledge constructors not as documenters of knowledge but as *carriers* of knowledge, culture and practices who fashion for us our apprehension of phenomenon and reality. Rather than some objective or universal entity representing some Truth out there, knowledge—in this representative capacity—constitutes for us the worlds we live in, and exemplifies for us, how and what we should be in our effective engagements with these worlds.

But this theoretical and methodological attentiveness to Self and Other informing my approach is also a textual one. As noted by Calás and Smircich (1999) in their assessment about the reconstructive possibilities engendered by postmodern and poststructuralist understandings of knowledge: "at the very moment the complicity of language in the constitution of knowledge becomes part of the "conversation," the "tone" of the conversation has to change" (: 652). From this standpoint, what I have sought to do—via the methodology of multivocality—is to provide a textual equivalent of being conceptually attentive to the multiplicities in voice, in identities and in social particularities which are seldom attended to in official stories on transfers.

It needs noting at this juncture that this attempt to write into the story both the author and the situated-ness of knowledge is becoming an increasingly prominent feature of writers bent on reconstituting a more accountable—and inclusive—form of *re*-presentation (e.g., Abu-Lughod, 1991; Anzaldua, 1987; Behar, 1996; Chio, 1993; Lal, 1996; Trinh T. Min-ha, 1989). From the contestations of workers to observations about the absence of cows and memory and history, to still more accounts about the living experience of being in the field and the process of knowing and representing, the textual "writing-ins" I forward as part of the story on transfers are intended to provide a more locally and representationally responsible account of both the phenomenon of transfers and the act of knowledge-gathering and knowledge-making itself.

As I seek to show in my personal account of the research and writing process in preceding chapters, knowledge and writing itself are always situated and coming from somewhere and intended for some end. In the specific instance of the personal text in chapter seven, what the account made possible is my ability to write-in and give voice to some of the struggles and issues informing my representation of the story that I tell here. From the acts of objectification to expectations governing authority and voice, and the myriad of progressive silences that mark these requirements and/or stages in the knowledge-making enterprise, what is given is an account of the author not just as agent of knowledge and power, but as a *subject* of knowledge and power. That is, the location of the author as a member of a larger academic and professional community who is vested with institutionalized and socialized expectations regarding norm and form, power and voice, science and legitimacy.

In terms of the actual physical text itself, this multivocality is represented by the use of a personal text that is inter-spliced into the official text. Strategically, what is significant here is the physical embedding or paralleling of the personal [or "unofficial"] into or with the official. In this way, what I seek to communicate is not just some of the social particularities [otherwise silenced or not attended to] organizing or locating the official story, but some of the multiplicities and effects that exist in tandem with—and alongside—the official story, so to speak.

Within the intersections of culture, knowledge and practice recounted here is a commentary about the problematic of researching phenomenon like knowledge transfers in the absence of conceptual frames and methodologies capable of recognizing and documenting the "interlocution" (Said, 1989) of knowledge by culture and practice, and culture by knowledge and practice. To this end, my objective in telling this story has not been so much to contribute to the further discovery of concepts, causalities, relations and so forth [i.e., the production of more facts (Smith, 1987)], but to illuminate and to problematize different ways, more historicized ways, more diversely-sensitive ways that can be accessed to reconstitute and understand processes that are *already* well presumed, researched and claimed.

RE-SITUATING MY APPROACH AND MY STORY

In sum, what I have sought to do in this project is to provide an alternative possibility for understanding, researching and representing Others. In doing so, I have drawn from the theoretical and methodological insights of scholars from elsewhere in the social sciences—anthropology, postcolonial studies, social theory, sociology—and extended it to the terrain of international and cross-cultural management. The writings of Smith (1987) on institutional ethnography, the focus on discourse by scholars deconstructing the development sciences (e.g., Crush, 1995; Escobar, 1994a, 1994b, 1988, 1984; Mueller, 1987a; Sachs, 1992), and the re-historicizing of relations between Self and Other by postcolonial historiographers (e.g., Appadurai, 1990; Chakrabarty, 1992; Chatterjee, 1993, 1987; Prakash, 1990; Harding, 1998; Radhakrishnan, 1996; Said, 1989) have been central in enabling me to formulate a framework and a methodology by which to configure the parameters of this project.

The project itself, however, is also situated within an already active scholarship in the field of organization studies that has been centrally engaged in bringing attention to the issues of knowledge and power, representation and voice that I call attention to here (e.g., Calás, 1987; Calás and Smircich, 1999, 1997, 1992; Cooper, 1989; Cooper and Burrell, 1988;

Jacobson and Jacques, 1997; Jacques, 1996; Knights, 1992; Putnam, 1996). Whether in reference to issues governing our understanding of globalization (Calás and Smircich, 1993; Mir, Calás and Smircich, 1999), organizations (e.g., Kilduff, 1993), race and diversity (e.g., Nkomo, 1992), gender (e.g., Calás, 1992; Calás and Jacques, 1988; Calás and Smircich, 1992; Martin, 1990), leadership (e.g., Calás, 1993; Calás and Smircich, 1991), and/or specific organizational practices like human resource management (e.g., Townley, 1994, 1993), these scholars have endeavored to provide alternative ways of understanding knowledge and the construction of knowledge which allows for a more accountable and reflexive engagement between knowledge and power, knowers and known.

According to Calás and Smircich in their assessment of postmodern theorizing's contribution to the terrain of organization studies, its significance lie "in the opportunities it has offered for reflecting upon the production of theory as a genre and as an institutional and cultural activity" (1999: 649). As they go on to note, the possibility of "viewing theory as a representational form" (: 649) has afforded the field with an opportunity to not only locate research and scholarship in terms of questions like "'for whom" and "for what'" (: 650), it has also allowed for the distinct possibility of "open[ing] "the margins" of organization studies to be "written" by and for others whose theoretical voices have seldom been represented in our scholarship" (: 650).

It is within these generalized and specific currents in our understandings and approach to knowledge that this book is situated. Within this context and this history, my central objective has been to illustrate and locate issues of representation and voice informing our knowledge and construction of knowledge in the international arena, focusing on Malaysia as an entry-point for locating some of these issues.

But this story and the issues it presents to us do not just deal with knowledge construction in the international and cross-cultural terrain. These issues and concerns also govern the here and the now of self and home, since the question or problem under contention here is less about the Other than it is about *the "Other" of our knowledge:* that is, the subjects and objects of our knowledge, and our engagements to and with them. In this sense, the questions I raise regarding issues of representation and voice, inclusion and exclusion, are equally applicable for considering where we are speaking from, and who we speak for, when we do and write organizational knowledge about the workplace, about workers and managers, organizational processes and managerial practices here in the West.

Put another way, the question is not so much about the possibility of representing those who are different, but how knowledge as we know and do it now is not attentive *to* difference—an indifference arising from and connected to its assumption that there is a fixedness to meanings, to interpretations and to identities on the one hand, and an endemic equality in power and voice on the other. These are points highlighted variously by authors noted above in their look at how issues of race and gender, worker and manager, have consistently been silenced by the representational practices we have relied on in order to know.

And so I return once more to the specific intersections and multiplicities in location, in intention and in history, which mark both this story and the storyteller I. Like the manager in Malaysia who finds himself "having to be two," the researcher I, in fashioning this story and this text, is actively involved in being a bridge between the multiple worlds that I am located in: child of a nation, member of a discipline and academic tradition, and researcher of homes-away-from-home. In doing so, I actively mediate between a local and a global, a here and a there, relying on procedures of knowledge and memory, representation and reflection in order to cultivate what I see as greater professional and personal accountability.

Yet even as I speak, nations like Malaysia are struggling still in the webs of control and power that I call attention to in this story, their economies decimated, financial policies forcibly opened-up to allow for easier capital inflows, and people finding themselves struggling yet again between power and voice, inclusion and exclusion. What are we coming to? And where are we headed for? The answers to these questions cannot be found in knowledge as we know it now since it is precisely this knowledge that accounts, and continues to account, for this disarray and this decimation. The preceding paragraph can thus be read as my own retreat from knowledge, as I seek to do "no harm" with my own work. And so I end.

Appendix A

Research Sites and Data Collection

I. MULTINATIONALS

1. *Infotech Incorporated*

• Formal and informal Interviews	• Vice-President and General Manager • Assistant General Manager • Heads of Departments and Strategic Business Units [manufacturing and support units] • Department Managers and Supervisors
• Observations	• Weekly staff meetings • Monthly communication event • Monthly quality excellence reviews • Outreach at local university
• Secondary Data	• Internal magazines and reports • Internal bulletins (concept papers, strategic evolutions in focus, samples of daily briefings, quality reviews) • Learning material

2. *Multitech Incorporated*

• Formal and informal Interviews	• General Manager/Managing Director • Head of the Commercial Department • Heads of (sub)-departments and staffers from the above (control systems and planning, TQM training and development, purchasing, finance, accounting, storage and inventory, human resources)

- Heads of 2 Manufacturing Business Units
- Observations
  - Heads of department meetings
  - HR weekly department meetings
  - TQM weekly and monthly communication events
  - New hires orientation
- Secondary Data
  - Internal magazines and bulletins
  - Company profiles, diagrams and reports (local, national and international)
  - Internal documents (organization chart, department objectives, production control charts and role descriptions, manufacturing management systems, manufacturing business unit objectives, TQM training statistics)
  - Copies of orientation and other more industry-wide videos, including tapings of internal events

II. STATE-INDUSTRY TRAINING CENTERS

1. Penang Skills Development Center (PSDC)

- Interview
  - Director
- Secondary Data
  - PSDC bulletin, report and updates
  - Program Catalogues

2. Selangor Human Resource Development Center

- Interview
  - Director
- Secondary Data
  - Flyer and basic description [little or none available since the center was being "set-up"]

III. STATE DEVELOPMENT AGENCIES

1. Penang Development Corporation (PDC)

- Interviews
  - Head of the Public Relations Unit
- Secondary Data
  - Annual reports, forecasts and handbook
  - Presentation materials

2. Selangor State Development Corporation (SDC)

• Interview	• Public Relations Manager
• Secondary Data	• Annual reports, corporate profiles and internal bulletin

IV. STATUTORY AND PARAPUBLIC AGENCIES

1. National Productivity Center (NPC)

• Interviews	• Director and Deputy Director General • Promotions Officers
• Observations	• Quality Control Circles Convention
• Secondary Data	• Annual reports and forecasts on the Center and national productivity • Public newsletters and bulletins • Sample of applications for Quality Awards • Internal bulletins and papers • Training material • Other information gleaned from the center's video library (quality references, tapings of previous visiting speakers)

2. Standards and Industrial Research Institute of Malaysia (SIRIM)

• Interviews	• Director and Deputy Director [units kept anonymous]
• Secondary Data	• Annual reports, profiles and bulletins • Quality training courses, planner and statistics • SIRIM LINK package • Video

V. FEDERAL AGENCIES & MINISTRIES

1. Malaysian Administration Modernization and Planning Unit (MAMPU)

• Interviews	• Director and Deputy Director [units kept anonymous]
• Secondary Data	• Internal magazine • Proposals, technical specifications and summary reports and outlines

(civil service link, integrated telecommunications network, quality management implementation)
- Publications: (1) The Civil Service of Malaysia. Towards Vision 2020, 1995, (2) Improvements and Developments in the Public Service, 1993, and (3) The Civil Service of Malaysia. A Paradigm Shift, 1994.

2. *Ministry of International Trade and Industry (MITI)*

- Interviews
    - Assistant Deputy Director [unit kept anonymous]
    - Official, Technology Transfer
- Secondary Data
    - Annual reports
    - Investment guidelines and projections
    - Literature (subcontract exchange, vendor development program, modernization & automation scheme, overview of SMI development)
    - Literature published in conjunction with other agencies (industrial digest, consultancy service scheme, product development & design scheme, market development scheme)

3. *Ministry of Education*

- Interviews
    - Director [unit kept anonymous]

VI. ACADEMIC INSTITUTIONS

1. *The University of Malaya (UM)*

2. *Universiti Kebangsaan Malaysia (UKM)*

3. *Universiti Sains Malaysia (USM)*

- Formal and Informal Interviews
    - A total of eleven individuals that include faculty members from three different schools to directors, deans, deputy deans and deputy vice chancellors were interviewed formally or informally. For reasons of confidentiality, I have elected to keep their specific designations anonymous.

- Secondary Data
  - Copy of papers from an international seminar sponsored by one of the university and government.

4. *Malaysian Institute of Management (MIM)*

- Interviews
  - Chief Executive Officer
- Observations
  - Seminar co-sponsored by the Konrad Adenauer Foundation (Germany) on "Malaysian Management Identity." Presenters and attendees consist of numerous representatives from the state and federal ministries, the private sector, parapublic sector and academia, both faculty and administrators.

    Included here are quite a number of representatives from the institutions and/or agencies noted in this Appendix.

Appendix B

Broad Objectives and Strategies for National Development in Malaysia

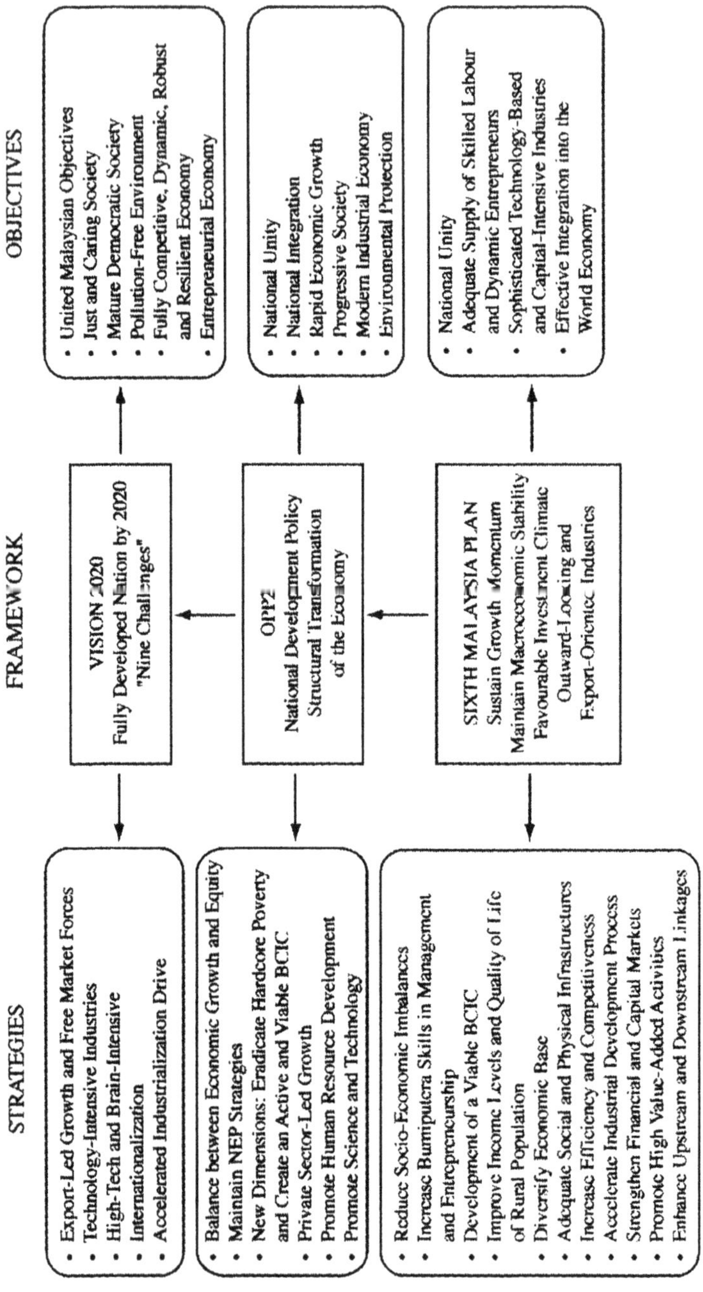

From [Figure 3.1] *Malaysia's Vision 2020: Understanding the Concept, Implications and Challenges* (p.70), by A. Sarji Abdul Hamid, 1995, Selangor, Malaysia: Pelanduk Publications (M) Sdn. Bhd.

Notes

NOTES TO CHAPTER ONE

1. Throughout the text, following the examples of writers like Irene Gendzier (1985), Adelle Mueller (1987a) and Arturo Escobar (1994a) in their analysis of modern knowledge construction practices in the social and development sciences, certain descriptors and concepts such as "industrialization," "development," "underdevelopment," "progress" and so forth are often deliberately capitalized to underline the problematic and contested nature of these concepts and institutions. Embedded within such a critique as was articulated by these authors is a growing recognition about the socially constructed nature of the modern episteme, reflecting as it does a form and practice of knowing which is at once cultured and institutionalized; usually involving complex imbrications between power and knowledge, knowledge and capital, science and administration, and so on.
2. The specific formulation here is a direct response to critiques raised by an increasing number of ethnographers of resistance and postcolonial historiographers of the Third World who highlight the often passive formulation of Third World and colonial subjects by Western anthropologists, scientists, historians, and so on. According to these authors, the Third World comes neither "empty of meanings" (e.g., Comaroff, 1985; Taussig, 1980), nor devoid of history (e.g., Chatterjee, 1993, 1987; Pandey, 1990).
3. See Stephen Gudeman (1986) and Arturo Escobar (1994a) for a discussion on economics as a social form and cultured construction of reality. For a somewhat different take on economics and culture, see also the edited volume by Don Lavoie (1990).
4. The specific construction here draws on Benedict Anderson's (1983) constitution of modern nation states as "imagined communities"; a formulation subsequently reformulated and expanded by Partha Chatterjee (1993) to include within this notion—particularly so in the context of the Third World—an understanding of nation states as consisting of simultaneous policies of inclusion and exclusion.

5. This process of self-discovery recalls Adelle Mueller's (1987a) contention, with specific reference to the Women in Development literature, that even when we seek to know and write about the Third World—or women in the Third World as the case may be—what we know are neither Third World nor women in the Third World, but the procedures of rule organizing our knowledge about these worlds.

NOTES TO CHAPTER TWO

1 For reasons of confidentiality and anonymity, the names of both multinationals have been changed. Concurrently, individuals from these and other institutions whom I interviewed and/or observed are referred to in this text, not by names but by locations (e.g., professional, institutional, social) in order to preserve their anonymity.

NOTES TO CHAPTER THREE

1. From Paul Rabinow. 1986. Representations are Social Facts: Modernity and PostModernity in Anthropology: 241. In James Clifford and George Marcus (Eds.) Writing Culture: The Poetics and Politics of Ethnography, Berkeley, CA: University of California Press.
2. According to Alain Touraine (1995) in his analysis on the rise and fall of modernity, there is a disjuncture between modernity as a mental revolution and the modernizations locating it as action and history: in effect, modernity, in the instance of "application" is fraught with resistances, appropriations and so on. It is a distinctly non-linear exercise and experiment directly associated with the constitution of modernity itself: namely, a rationalist purview intent and premised on the possibility of separating understanding and experience, apprehension and understanding, history and present and so on.
3. For examples of dependency theories and their analysis of the ingrained structural dependencies created by interconnections between western capital and industrialization, see Paul Baran (1957), Fernando H. Cardoso and Enzo Faletto (1979), Andre Gunder Frank (1967), and Celso Furtado (1978, 1970). For accounts of the uneven costs and effects borne by constituents, particularly women in the Third World, as a result of multinationals' global sourcing activities and governments' export industrialization strategies, see Maria Patricia Fernández-Kelly (1994, 1983); Annette Fuentes and Barbara Ehrenreich (1983), Rachel Grossman (1978/1979), Susan Joekes (1987), Linda Lim (1985, 1983, 1981), and Aihwa Ong (1993, 1987). Analysis of the reductionist violence [scientific, economic] and marginalization of local histories, identities, forms and alternatives resulting from modern development interventions can be seen in most of the development critiques (e.g., Shiva et. al.) noted later in the chapter. See also Jonathan Crush (1995), Arturo Escobar (1994b, 1988, 1984), Arturo Escobar and Sonia Alvarez (1992), Rajni Kothari (1988, 1987), and Stephen Marglin and Frédérique Appfel-Marglin (1990). For more explicit analysis of the connections between colonialism—discursive, material, cultural, political, psychological—and the Third World, refer to

postcolonial writings (e.g., Chatterjee et. al.) already listed. See also Homi K. Bhabha (1994), Frantz Fanon (1967), Albert Memmi (1967), Lata Mani (1989), Chandra T. Mohanty, Ann Russo and Lourdes Torres (1991), Gayatri Spivak (1990, 1987) and Edward Said (1983, 1979).

NOTES TO CHAPTER FOUR

1. Bumiputera, translated as "true sons of the soil', is a social and policy category referring to Malays and other indigenous groups in Malaysia. Since the beginning of their administration, the British had regarded the Malays as the sovereign rulers of the land, while the migrant communities, whose presence in the nation is due in no small sense to the same administration's importation of migrant workers to work in the plantations, mines and civil service, were regarded as temporary "guests" (Khoo, 1995; Crouch, 1996). The category is institutionalized into the constitution and independence principles of the nation, conferring on the community the privileges accorded their status as the "definitive people" (Mohamad, 1982) of the land.
2. A coalition between the United Malays National Organization (UMNO), the Malayan Chinese Association (MCA), and the Malayan Indian Congress (MIC), the Alliance was an attempt by UMNO leaders to recognize and deal with the pluralistic feature of Malayan society via the formulation of political "alliances" with moderate non-Malay leaders from the two major "minority" groups in the country. Initially formed in 1952, the Alliance won all national elections between 1955 and 1969 (Crouch, 1996).
3. British administration had set out early on to "create an English-educated administrative class recruited from the Malay aristocracy", who tended to fill the "highest echelons of the bureaucracy". (Crouch, 1996: 16–17). A relatively homogeneous elite, leaders in the bureaucracy, the armed forces, the judiciary and so on tended to have "similar values, orientations, educational backgrounds, and life-styles" (: 19); the majority of whom were conservative and pro-British.
4. While per capita income had increased by 25% between 1957 and 1970, there still existed big disparities between communities; it is estimated that while the income of the top 10% of the population actually increased by over 50%, the average income of the bottom 40% actually declined by 13%; with intercommunity differential approximately on a two-to-one basis between the Chinese and the Malay communities (Crouch, 1996: 21).
5. Policy-wise, the NEP is defined by a two-pronged approach involving (1) the eradication of poverty irrespective of race, and (2) the restructuring of society to reduce the identification of race with economic function (Government of Malaysia, 1971). Procedurally, the former is targeted mainly at reducing poverty in the rural areas where Malays predominate, while the latter involves a massive reconstitution to the social, educational and economic sectors in the nation. Included here is the setting of employment targets in all sectors of the economy, and the restructuring of ownership in share capital such that by the end of the NEP's tenure, not only will Malay representation in modern sectors of the economy better reflect their "racial composition" in the nation, they will also own and/or

operate at least 30% of the nation's "productive wealth" (Government of Malaysia, 1975: 75–76). It needs noting at this juncture that while the NEP's focus on restructuring ownership of share capital often involved a variety of preferential treatment for Malays and Bumiputeras at the expense of the non-Malay community (e.g., access to licenses, concessions, contracts, credit), it's main target for restructuring is foreign, primarily British, capital (Crouch, 1996). As Crouch notes: "the common perception that the modern economy was dominated by the Chinese masked the even more dominant role of foreigners" (: 26). According to him, over 63% of share capital in West Malaysia was in the hands of foreign capital in 1970. Thus, the pain of restructuring for non-Malays was to be partially ameliorated by reducing foreign ownership; and the goal of increasing Malay ownership achieved not so much by reducing non-Malays' share, but by restricting and regulating foreign ownership.

6. A constitution of the NEP first brought to my attention by a colleague in Malaysia who was quoting the words and sentiment of a local Malay journalist and writer, Rehman Rashid.

NOTES TO CHAPTER FIVE

1. From Mahathir Mohamad. 1993b. Malaysia: The Way Forward: 415. In Ahmad Sarji Abdul Hamid's (Ed.) Malaysia's Vision 2020: Understanding the Concept, Implications and Challenges. Petaling Jaya, Malaysia: Pelanduk Publications.
2. A reminder to note at this point, how such constructions of subjects in child-like terms has been criticized by organizational psychologists like Chris Argyris (1957). According to him, these child-like constitution of workers not only frustrates their ability to grow and take on more responsibility, they are unreflective about how the structural and organizational exigencies of the modern workplace (e.g., rules and standards of procedures) are actually producing and organizing the activities and capacities of these workers. Although it needs noting at this juncture that Argyris' critique is forged still within the vagaries of what makes for, or maybe more appropriately given his critique, what allows for a good organizational worker.
3. During the early and mid-1980s, a series of seminars and workshops on productivity forming part of the NPC's "Productivity Movement" [launched in 1982] was held throughout the country which led to the start of its productivity measurement activities. The organizational framework (NPC, 1994, Box 1: ii) geared towards this objective of compiling and analyzing productivity statistics, put in place in 1990 and 1994, include seven Directorates for Productivity Measurement and Research, and a Consultative Panel for Research. Set up "to provide current data on productivity and quality, to disseminate productivity concepts, systems and methodologies, and to advise the government on . . . productivity and quality" (NPC, 1994: ii), each Directorate is responsible for certain sub-sectors in the manufacturing and services industries. Concurrently, the

Consultative Panel, composed of members from industry, the public sector, institutions of higher learning and unions, is intended to the assist the NPC in "formulating and implementing strategies for . . . develop[ing] productivity statistics and networks," "advis[ing] . . . on approaches to measuring and forecasting productivity," "developing linkages between productivity and performance measures and other economic indicators" and "assessing and monitoring productivity and quality development in the country" (: ii). Funded by the ILO, the United Nations Development Programme (UNDP) and the Asian Productivity Organization (APO), experts from these institutions were brought in to aid the NPC in developing productivity and quality measures of [economic and development] performance.

4. As a statutory body, all policy for running the agency is undertaken by SIRIM while financial support comes from the government, although with its "corporatization" in January, 1996, the Institute is to all intents and purposes, an independent financial entity [albeit one where the government is its sole shareholder]. Under this new arrangement, government agencies are themselves required to reimburse to the Institute a sum on par with the market value of whatever services are rendered (e.g., contract research, technical assessment). Structurally, it is composed of a corporate services unit that supports three technical sectors: standards and quality, research and industrial technology, and technology transfers. This last is specifically targeted at local SMIs.
5. By the time of the publication of its 1994 annual report, SIRIM is the official certification agent for 12 certification bodies. In 1994 alone, the Institute—as part of its objective to "assist local industry [in] penetrat[ing] global markets" (SIRIM, 1994: 50)—was able to sign seven more memoranda of understandings with a variety of institutions around the world. In its capacity as a certification agent for ISO 9000, the Institute registered a total of 503 companies between 1990 and 1994, with the bulk of these [455] approved between 1992 and 1994 (: 41).
6. Multiple connections are also being forged between the state and universities [often enforced via the shifting of resources to industrial research], or multinationals and universities (e.g., developing or tailoring upper degree programs to meet the needs of companies, researching topics or concerns requested by companies). Often cultivated under the auspices of deregulation and HRD, these efforts are encouraged by state planners in close connection with their intent to corporatize or privatize local universities, as well as to increase the market-relevancy and technological base of the nation. An interesting variant on this took on the form of a proposal tabled by one of the multinationals I observed to aid in the HRD and educational imperative in the nation. The focus: cultivation of a long distance education alternative via live telecasts of classes held at a Western university broadcasted to a local university. Costs for the satellite and studio were paid for by the multinational. The state provided and approved the use of a local university and its facilities as the site for such an activity, with the understanding that the setting up of [subsequent?] facilities and purchases of equipment and so forth would be paid for by the

state. For the government, it was a way of expediting the shortage of training and educational infrastructure in the nation. On the part of the MNC, it was a way of ensuring the availability and supply of required manpower in targeted areas like engineering [and management].

7. As an example, NPC training efforts in 1995 alone included a plethora of foreign experts from universities in Japan and the US. Noted earlier, reliance on foreign expertise from the ILO, UNDP and APO has enabled the NPC itself to fulfill its development, information gathering and assessment aims in regard to the development of national productivity measures and evaluation. On SIRIM's part, a total of 22 experts, technology consultants and research fellows were recruited in 1994 [under the Industrial Technology Development and the UNDP Programs] to aid its numerous technology groups and technical areas with "hands-on training and consultancy much needed in building up [its] capability and experience in the [designated] areas" (SIRIM, 1994: 51).

NOTES TO CHAPTER SIX

1. For an interesting study on how resistance is forged within the confines of a highly regulated social regime, refer to Tan Sooi Beng's (1992) "Counterpoints in the Performing Arts in Malaysia". In this article, she refers attention to implicit and explicit critiques of Malaysian society—social, political, bureaucratic—that are currently emerging in a variety of popular culture forms, whether derived from more traditional, modern or some amalgamation of these genres. A key intent behind her article is to assess how social critiques are, or can be forged even in the midst of a social and cultural policy that sees the "active production of culture" by state authorities as part and parcel of national security.

Bibliography

Abu-Lughod, L. 1991. Writing Against Culture. In R. Fox (Ed.) *Recapturing Anthropology: Working in the Present.* Santa Fe, NM: School of American Research.

Abu-Lughod, L. 1986. *Veiled Sentiments: Honor and Poetry in a Bedouin Society.* Berkeley, CA: University of California Press.

Achebe, C. 1976. Colonialist Criticism. *In Morning Yet on Creation Day, Essays by Chinua Achebe:* 3–24. Garden City, NY: Anchor Books.

Adler, N. J. 1984. Understanding the Ways of Understanding: Cross-Cultural Management Reviewed. In R. N. Farmer (Ed.) *Advances in International Comparative Management,* 1: 31–67. Greenwich, CT: JAI Press.

Adler, N. J. 1983. Cross-Cultural Management Research: The Ostrich and the Trend. *Academy of Management Review,* 8 (2): 226–232.

Adler, N. J. & Graham, J. L. 1989. Cross-Cultural Interaction: The International Comparison Fallacy? *Journal of International Business Studies,* Fall: 515–537.

Adler, N. J. & Jelinek M. 1986. Is Organization Culture "Culture" Bound? *Human Resource Management,* 25: 73–90.

Adler, N. J., Campbell, N. & Laurent, A. 1989. In Search of Appropriate Methodology: From Outside the People's Republic of China Looking In. *Journal of International Business Studies,* Spring: 61–74.

Ali, A. 1992. *Malaysia's Industrialization: The Quest for Technology.* Singapore: Oxford University Press.

Amin, S. 1990. *Maldevelopment.* London: Zed Books.

Anderson, B. 1983. *Imagined Communities.* London: Verso Books.

Anzaldua, G. 1987. *Borderlands/La Frontera: The New Mestiza.* San Francisco: Aunt Lute Books.

Apffel-Marglin, F. and Marglin, S. 1990. *Dominating Knowledge: Development, Culture and Resistance.* Oxford: Clarendon Press.

Appadurai, A. 1990. Disjuncture and Difference in the Global Cultural Economy. *Public Culture,* 2 (2): 1–24.

Argyris, C. 1957. *Personality and Organization.* New York: Harper.

Baran, P. 1957. *The Political Economy of Growth.* New York: Monthly Review Press.

Barnard, C. 1938. *The Functions of the Executive.* Cambridge, MA: MIT Press.

Bartlett, C. A. & Ghoshal, S. 1995. *Transnational Management.* Boston: Irwin McGraw-Hill.

Bartlett, C. A. & Ghoshal, S. 1989. *Managing Across Cultures: The Transnational Solution.* Boston: Harvard University School Press.

Bartlett, C. A. & Ghoshal, S. 1988. Creation, Adoption, and Diffusion of Innovation by Subsidiaries of Multinational Corporations. *Journal of International Business Studies,* 19: 365–388.

Bartlett, C. A. & Ghoshal, S. 1987a. Managing Across Borders: New Organizational Responses. *Sloan Management Review,* Fall: 43–53.

Bartlett, C. A. & Ghoshal, S. 1987b. Managing Across Borders: New Strategic Requirements. *Sloan Management Review,* Summer: 7–17.

Behar, R. 1996. *The Vulnerable Observer: Anthropology That Breaks Your Heart.* Boston: Beacon Press.

Berthoud, G. 1992. Market. In W. Sachs (Ed.) *The Development Dictionary. A Guide to Knowledge as Power:* 70–87. London: Zed Books.

Bhabha, H. K. 1994. *The Location of Culture.* London and New York: Routledge.

Bhabha, H. K. (Ed.) 1990. *Nation and Narration.* New York: Routledge.

Bourdieu, P. 1977. *Outline of a Theory of Practice.* Cambridge: Cambridge University Press.

Boyacigiller, N. A. & Adler, N. J. 1991. The Parochial Dinosaur: Organizational Science in a Global Context. *Academy of Management Review,* 16 (2): 262–281.

Bremer, J. A. 1984. Building Institutional Capacity for Policy Analysis: An Alternative Approach to Sustainability. *Public Administration and Development,* 4: 1–13.

Buckley, P. J. 1991. The Frontiers of International Business Research. *Management International Review* [Special Issue]: 2–22.

Buckley, P. J. & Casson, M. 1988. A Theory of Cooperation in International Business. In F. J. Contractor & P. Lorange (Eds.) *Cooperative Strategies in International Business:* 31–54. Lexington, MA: Lexington Books.

Burrell, G., Calás, M. B., Morgan, G., Reed, M. & Smircich, L. (Eds.) 2000. [Special Issue]. *Organization,* 7 (3).

Burrell, G. & Morgan, G. 1979. *Sociological Paradigms and Organizational Analysis. Elements of the Sociology of Corporate Life.* Portsmouth, NH: Heinemann Educational Books Inc.

Calás, M. B. 1993. Deconstructing Charismatic Leadership: Re-reading Weber from the Darker Side. *Leadership Quarterly,* 4: 305–328.

Calás, M. B. 1992. An/Other Silent Voice? Representing "Hispanic Woman" in Organizational Texts. In A. J. Mills & P. Tancred (Eds.) *Gendering Organizational Analysis:* 201–221. Newbury Park, CA: Sage Publications.

Calás, M. B. 1987. *Organization Science/Fiction: The Postmodern in the Management Disciplines.* Unpublished doctoral dissertation. School of Management, University of Massachusetts at Amherst.

Calás, M. B & Jacques, R. 1988. Diversity or Conformity?: Research by Women on Women in Organizations. Paper presented at the Seventh International Conference on Women and Organizations, Long Beach, California.

Calás, M. B & Smircich, L. 1999. Past Postmodernism?: Reflections and Tentative Directions. *Academy of Management Review,* 24 (4): 649–671.

Calás, M. B & Smircich, L. 1997. *Postmodern Management Theory.* Aldershot, England: Ashgate/Dartmouth.

Calás, M. B & Smircich, L. 1993. Dangerous Liaisons: The Feminine-in-Management Meets Globalization. *Business Horizons,* 36 (2): 71–81.

Calás, M. B & Smircich, L. 1992. Re-writing Gender into Organizational Theorizing: Directions from Feminist Perspectives. In M. Reed & M. Hughes (Eds.) *Rethinking Organizations:* 227–253. London: Sage.

Calás, M. B & Smircich, L. 1991. Voicing Seduction to Silence Leadership. *Organization Studies,* 12: 567–602.

Cardoso, F. H. & Faletto E. 1979. *Dependency and Development in Latin America.* Berkeley: University of California Press.

Cavendish, R. 1982. *Women on the Line.* London: Routledge and Kegan Paul.

Chakrabarty, D. 1992. Postcoloniality and the Artifice of History: Who Speaks for "Indian" Pasts? *Representations,* 37: 1–26.

Chatterjee, P. 1993. *The Nation and Its Fragments.* Princeton, NJ: Princeton University Press.

Chatterjee, P. 1988. For an Indian History of Peasant Struggle. *Social Scientist,* 16 (11): 3–17.

Chatterjee, P. 1987. *Nationalist Thought and the Colonial World: a Derivative Discourse.* London: Zed Books Ltd.

Chatterjee, P. 1985. Modes of Power: Some Clarifications. *Social Scientist,* 13 (2): 52–63.

Chio, V. 1993. "Unbounding" Organizational Analysis in the International Arena: Third World Feminism and the Politics of Representation. Paper presented at the Annual Meeting of the National Academy of Management, Atlanta, Georgia.

Christelow, D.B. 1987. International Joint Ventures: How Important are They? *Columbia Journal of World Business,* 22 (2): 7–13.

Clark, P. A. 1982. Temporal Innovations and Time Structuring in Large Organizations. In J. T. Fraser, N. Lawrence & D. Park (Eds.) *The Study of Time.* Volume 3. New York: Springer-Verlag.

Clifford, J. 1988. On Ethnographic Authority. In J. Clifford (Ed.) *The Predicament of Culture: Twentieth Century Ethnography, Literature, and Art:* 21–54. Cambridge, MA: Harvard University Press.

Clifford, J. 1986. Introduction: Partial Truths. In J. Clifford & G. Marcus (Eds.) Writing Culture: *The Poetics and Politics of Ethnography.* Berkeley: University of California Press.

Clifford, J. & Marcus, G. E. (Eds.) 1986. *Writing Culture: The Poetics and Politics of Ethnography.* Berkeley: University of California Press.

Comaroff, J. 1985. *Body of Power, Spirit of Resistance: The Culture and History of a South African People.* Chicago: The University of Chicago Press.

Cooper, R. 1989. Modernism, Postmodernism and Organizational Analysis 3: The Contribution of Jacques Derrida. *Organization Studies,* 10: 479–502.

Cooper, R. & Burrell, G. 1988. Modernism, Postmodernism and Organizational Analysis: An Introduction. *Organizational Analysis,* 9: 91–112.

Crouch, H. 1996. *Government and Society in Malaysia.* Ithaca, NY: Cornell University Press.

Crush, J. (Ed.) 1995. *Power of Development.* London: Routledge.

Davies, D. G. 1980. Human Development in Sub-Saharan Africa. In *Poverty and the Development of Human Resources: Regional Perspectives*. Washington, DC: World Bank Staff Working Papers, No. 406.
Ditton, J. 1979. Baking Time. *Sociological Review,* 27: 157–167.
Doktor, R., Kawase, T. & Haig, J. H. 1986. Culture as a Constraint on Productivity. *International Studies of Management and Organization,* 15 (3–4): 8–16.
Doktor, R., Tung, R. L. & Von Glinow, M. A. 1991a. Incorporating International Dimensions in Management Theory Building. *Academy of Management Review,* 16 (2): 259–261.
Doktor, R., Tung, R. L. & Von Glinow, M. A. 1991b. Future Directions for Management Theory Development. *Academy of Management Review,* 16 (2): 262–265.
Dror, Y. 1984. Facing Momentous Choices. *International Review of Administrative Sciences,* 2: 97–106.
Dunning, J. H. 1993. *Multinational Enterprises and the Global Economy.* Wokingham Berks, England: Addison-Wesley.
Dunning, J. H. 1989. The Study of International Business. *Journal of International Business Studies,* Fall: 411–436.
Dwyer, K. 1987. *Moroccan Dialogues: Anthropology in Question.* Prospect Heights, IL: Waveland Press Inc.
Economist. November 27, 1999. Trust Me, I'm the Doctor: 39 -40.
Emmerji, L. 1992. Globalization, Regionalization and World Trade. *Columbia Journal of World Business,* 27 (2): 7–13.
Edwards, S. 1995. *Crisis and Reform in Latin America: From Despair to Hope.* New York: Oxford University Press.
Erden, D. 1988. Impact of Multinational Companies on Host Countries: Executive Training Programs. *Management International Review,* 28 (3): 39–47.
Escobar, A. 1994a. Welcome to Cyberia: Notes on the Anthropology of Cyberculture. *Current Anthropology,* 35 (3): 211–223.
Escobar, A. 1994b. *The Development Encounter: The Making and Un-Making of the Third World, 1945–1992.* Princeton, NJ: Princeton University Press.
Escobar, A. 1992. Reflections on "Development": Grassroots Approaches and Alternative Politics in the Third World. *Futures,* 24 (5): 411–436.
Escobar, A. 1991. Social Science Knowledge and Development Discourse. Presentation prepared for the Social Science Research Council (SSRC) meeting in Berkeley, California.
Escobar, A. 1988. Power and Visibility: Development and the Invention and Management of the Third World. *Cultural Anthropology,* 3 (4): 428–443.
Escobar, A. 1984. Discourse and Power in Development: Michel Foucault and the Relevance of His Work to the Third World. *Alternatives,* 10 (3): 377–400.
Escobar, A. & Alvarez, S., (Eds.) 1992. *The Making of Social Movements in Latin America.* Boulder, CO: Westview Press.
Esman, M. J. 1991. *Management Dimensions of Development: Perspectives and Strategies.* West Hartford, CT: Kumarian Press.
Esteva, G. 1992. Development. In W. Sachs (Ed.) *The Development Dictionary: A Guide to Knowledge as Power:* 6–25. London: Zed Books.
Esteva, G. 1987. Regenerating People's Space. *Alternatives,* 12 (1): 125–152.

Fabian, J. 1983. *Time and the Other: How Anthropology Makes its Object*. New York: Columbia University Press.
Fanon, F. 1967. *Black Skin, White Masks*. New York: Grove Press.
Fayol, H. 1949. *General and Industrial Management*. New York: Pitman.
Ferguson, J. 1990. *The Anti-Politics Machine: "Development," Depoliticization and Bureaucratic Power in Lesotho*. Cambridge: Cambridge University Press.
Fernandez-Kelly, M. P. 1994. Making Sense of Gender in the World Economy: Focus on Latin America. *Organization*, 1 (2): 249–176.
Fernandez-Kelly, M. P. 1983. *For We are Sold, I and My People: Women and Industry in Mexico's Frontier*. Albany, NY: State University of New York Press.
Foucault, M. 1980. *The History of Sexuality, Volume 1: An Introduction*. New York: Vintage.
Foucault, M. 1979. *Discipline and Punish*. New York: Vintage Books.
Frank, A. G. 1967. The Development of Underdevelopment. *Monthly Review Press*, 18 (4): 17–31.
Frankenstein, J. 1986. Trends in Chinese Business Practice: Changes in the Beijing Wind. *California Management Review*, 29 (1): 148–160.
Fuentes, A. and Ehrenreich, B. 1983. *Women in the Global Economy*. Boston: South End Press.
Furtado, C. 1978. *Economic Development of Latin America: Historical Background and Contemporary Problems*. New York: Cambridge University Press.
Furtado, C. 1970. *Obstacles to Development in Latin America*. Garden City, NY: Anchor Books.
Garfinkel, H. 1967. *Studies in Ethnomethodology*. Englewood Cliffs, NJ: Prentice-Hall.
Geertz, C. 1983. "From the Native's Point of View": On the Nature of Anthropological Understanding. In C. Geertz (Ed.) Local Knowledge: *Further Essays in Interpretive Anthropology*: 55–70. New York: Basic Books.
Geertz, C. 1973. Thick Description: Toward an Interpretive Theory of Culture. In C. Geertz (Ed.) *The Interpretation of Cultures*: 3–30. New York: Basic Books.
Gendzier, I. 1985. *Managing Political Change: Social Scientists and the Third World*. Boulder, CO: Westview Press.
Gherardi, S. & Nicolini, D. 2000. To Transfer is to Transform: The Circulation of Safety Knowledge. *Organization*, 7 (2): 329–348.
Goes, J. B. and Park, S. H. 1997. Interorganizational Links and Innovation: The Case of Hospital Services. *Academy of Management Journal*, 40 (3): 673–696.
Goffman, E. 1959. *The Presentation of Self in Everyday Life*. New York: Doubleday/Anchor.
Government of Malaysia. 1995a. *Malaysia International Trade and Industry Report 1994*. Ministry of International Trade and Industry. Kuala Lumpur, Selangor.
Government of Malaysia. 1995b. *Economic Report 1995/1996*. Ministry of Finance. Kuala Lumpur, Malaysia: Percetakan Nasional Malaysia Berhad.
Government of Malaysia. 1994. *Economic Report 1994/95*. Ministry of Finance. Kuala Lumpur, Malaysia: Percetakan Nasional Malaysia Berhad.
Government of Malaysia. 1991. *Sixth Malaysia Plan: 1991–1995*. Kuala Lumpur, Selangor.
Government of Malaysia. 1981. *Fourth Malaysia Plan: 1981–1985*. Kuala Lumpur, Selangor.

Government of Malaysia. 1975. *Third Malaysia Plan: 1976–1980*. Kuala Lumpur, Selangor.

Government of Malaysia. 1971. *Second Malaysia Plan: 1971–1975*. Kuala Lumpur, Selangor.

Grant, R. M. 1991. Porter's "Competitive Advantage of Nations": An Assessment. *Strategic Management Journal*, 12: 535–548.

Grossman, R. 1978/79. Women's Place in the Integrated Circuit. Joint issue of *Southeast Asia Chronicle*, 66, and *Pacific Research*, 9 (5–6): 2–16.

Gudeman, S. 1986. *Economics as Culture*. London: Routledge and Kegan Hall.

Guha, R. 1983. *Elementary Aspects of Peasant Insurgency in Colonial India*. New Delhi, India: Oxford University Press.

Gupta, A. K. & Govindarajan, V. 1991. Knowledge Flows and the Structure of Control within Multinational Corporations. *Academy of Management Review*, 16 (4): 768–792.

Hamel, G., Doz, L. Y. & Pralahad, C. K. 1989. Collaborate with your Competitors and Win. *Harvard Business Review*, 67 (1): 133–139.

Hamel, G. & Pralahad, C. K. 1985. Do You Really Have a Global Strategy? *Harvard Business Review*, July/August: 139–148.

Haraway, D. 1988. Situated Knowledges: The Science Question in Feminism and the Priviledge of Partial Perspective. *Feminist Studies*, 14 (3): 575–599.

Harding, S. G. 1998. *Is Science multi-cultural?: Postcolonialisms, Feminisms, and Epistemologies*. Bloomington, IN: Indiana University Press.

Harpaz, I. 1990. The Importance of Work Goals: An International Perspective. *Journal of International Business Studies*, 21 (1): 75–93.

Hassard, J. 1999. Images of Time in Work and Organization. In S. R. Clegg & C. Hardy (Eds.) *Studying Organization: Theory and Method*: 327–344. London: Sage.

Harvey, D. 1989. *The Condition of Postmodernity*. Oxford: Basil Blackwell.

Henley, J. S. & Nyaw, M. K. 1986. Introducing Market Forces into Managerial Decision-Making in Chinese Industrial Enterprises. *Journal of Management Studies*, 23 (6): 635–655.

Herzberg, F. 1966. *Work and the Nature of Man*. Cleveland, OH: World Publishing.

Hirst, P. & Thompson, G. 1994. Globalization, Foreign Direct Investment and International Economic Governance. *Organization*, 1 (2): 277–303.

Hitt, M.A., Keats, B. W. & DeMarie, S. M. 1998. Navigating in the New Competitive Landscape: Building Strategic Flexibility and Competitive Advantage in the Twenty-First Century. *Academy of Management Executive*, 12 (4): 22–42.

Hofstede, G. 1992. Cultural Constraints in Management Theories. *Academy of Management Executive*, 7 (1): 81–94.

Hofstede, G. 1987. The Cultural Relativity of Organizational Practices and Theories. In A. Dymza & J. Vamberly (Eds.) *International Business Knowledge*: 309–323. New York: Praeger.

Hofstede, G. 1985. The Interaction Between National and Organizational Value Systems. *Journal of Management Studies*, 22 (4): 347–357.

Hofstede, G., Bond, M. H. & Luk, C. L. 1993. Individual Perceptions of Organizational Cultures: A Methodological Treatise on Levels of Analysis. *Organization Studies*, 14 (4): 483–503.

Hope, K. R. 1982. Improving Public Enterprise Management in Developing Countries. *Journal of General Management,* 7 (3): 72–85.

Hout, T., Porter, M. & Rudden, E. 1982. How Global Companies Win Out. *Harvard Business Review,* September/October: 98–108.

Hua, W. Y. 1983. *Class and Communalism in Malaysia.* London: Zed Books.

Husin Ali, S. (Ed.) 1984. *Ethnicity, Class and Development: Malaysia.* Kuala Lumpur, Malaysia: Persatuan Sains Social Malaysia.

Ismail, M. N. 1995. *Transnational Corporations and Economic Development.* Kuala Lumpur, Malaysia: University of Malaya Press.

Israel, A. 1987. *Institutional Development: Incentives to Performance.* A World Bank Publication. Baltimore, MD: John Hopkins University Press.

Jacobson, S. & Jacques, R. 1997. Destabilizing the Field: Poststructuralist Knowledge-Making Strategies in a Postmodern Era. *Journal of Management Inquiry,* 6: 42–59.

Jacques, R. 1996. *Manufacturing the Employee.* London: Sage.

Jay, M. 1988. Scopic Regimes of Modernity. In H. Foster (Ed.) *Vision and Visuality:* 3–23. Seattle, WA: Bay Press.

Jelinek, M. & Adler, N. J. 1988. Women: World Class Managers for Global Competition. *Academy of Management Executive,* February: 7–19.

Jesudasen, J. 1988. *Ethnicity and the Economy: The State, Chinese Business and Multinationals in Malaysia.* Singapore: Oxford University Press.

Joekes, S. 1987. *Women in the World Economy: An INSTRAW Study.* New York: Oxford University Press.

Jomo, K. S. 1993. The Way Forward?: The Political Economy of Development Policy Reform in Malaysia. An inaugural lecture delivered at the University of Malaya, Faculty of Economics and Administration, University of Malaya, Kuala Lumpur, Malaysia.

Jomo, K. S. 1986. *A Question of Class: Capital, the State and Uneven Development in Malaysia.* Singapore: Oxford University Press.

Jones, M. L. 1989. Management Development: An African Focus. *International Studies of Management and Organization,* 19 (1): 74–90.

Jorgensen, J. G. 1979. Cross-Cultural Comparisons. *Annual Review of Anthropology,* 8: 309–331.

Jorgensen, J. J., Hafsi, T. & Kiggundu, M. N. 1986. Towards a Markets Imperfections Theory of Organizational Structure in Developing Countries. *Journal of Management Studies,* 23 (4): 417 -441.

Kahn, J. S. & Loh, F. 1993. *Fragmented Vision: Culture and Politics in Contemporary Malaysia.* Sydney: Allen and Unwin.

Kane, M. and Ricks, D. 1989. The Impact of Transborder Data Flow Regulations on Large US-based Corporations. *Columbia Journal of World Business,* 24: 23–29.

Kane, M. and Ricks, D. 1988. Is Transnational Data Flow Regulation a Problem? *Journal of International Business Studies,* 19 (3): 477–482.

Kanungo, R. N. 1983. Work Alienation: A Pancultural Approach. *International Studies of Management and Organization,* 13 (1–2): 119–138.

Kanungo, R. N. 1982. *Work Alienation: An Integrative Approach.* New York: Praeger.

Kanungo, R. N. & Wright, R. W. 1983. A Cross-Cultural Comparative Study of Managerial Job Attitudes. *Journal of International Business Studies,* Fall:115–129.

Katz, D. & Kahn, R. L. 1966. *The Social Psychology of Organizations*. New York: John Wiley.

Kedia, B. L. & Bhagat, R. S. 1988. Cultural Constraints on Transfer of Technology Across Nations: Implications for Research on International and Comparative Management. *Academy of Management Review,* 13 (4): 559–571.

Khandwalla, P. N. 1989. OB for Social Development. *International Studies of Management and Organization,* 18 (4): 6–44.

Khandwalla, P. N. 1984. Some Lessons for the Management of Public Enterprises. *International Studies of Management and Organization,* 14 (2–3): 167–196.

Khoo, B. T. 1995. *Paradoxes of Mahathirism: An Intellectual Biography of Mahathir Mohamad.* Kuala Lumpur, Malaysia: Oxford University Press.

Khoo, K. J. 1992. The Grand Vision: Mahathir and Modernisation. In J. S. Kahn & F. Loh (Eds.) *Fragmented Vision: Culture and Politics in Contemporary Malaysia:* 44–76. Sydney: Allen and Unwin.

Kiggundu, M. N. 1989. *Managing Organizations in Developing Countries: An Operational and Strategic Approach.* Hartford, CT: Kumarian Press.

Kiggundu, M. N. 1986. Limitations to the Application of Sociotechnical Systems in Developing Countries. *Journal of Applied Behavioral Science,* 22 (3): 341–353.

Kilduff, M. 1993. Deconstructing Organizations. *Academy of Management Review,* 18: 13–31.

Knights, D. 1992. Changing Spaces: The Disruptive Impact of a New Epistemological Location for the Study of Management. *Academy of Management Review,* 17: 514–536.

Kondo, D. 1997. About Face: *Performing Race in Fashion and Theater.* New York: Routledge.

Kotabe, M. & Omura, G. S. 1989. Sourcing Strategies of European and Japanese Multinationals: A Comparison. Journal of International Business Studies, Spring: 113–130.

Kothari, R. 1988. Rethinking Development: In Search of Humane Alternatives. Delhi, India: Ajanta.

Kothari, R. 1987. On Humane Governance. Alternatives, 12 (3): 277–290.

Kumar, U. & Singh. K. K. 1978. The Interpersonal Construct System of an Indian Manager: A Determinant of Organizational Behavior. Management International Review, 18 (1): 49–59.

Lakoff, G. & Johnson, M. 1980. Metaphors We Live By. Chicago: University of Chicago Press.

Lal, J. 1996. Situating Locations: The Politics of Self, Identity, and "Other." In D. L. Wolf (Ed.) Feminist Dilemmas in Fieldwork: 185–214. Boulder, CO: Westview Press.

Lane, H. W., DiStefano, J. J. & Maznevski, M. L. 1997. International Management Behavior. Third Edition. Cambridge, MA: Basil Blackwell.

Lavoie, D. 1990. Economics and Hermeneutics. London: Routledge.

Law, J. 1999. Notes on the Theory of the Actor-Network: Ordering, Strategy, and Heterogeneity." In Warrick Organizational Behaviour Staff (Eds.) Organizational Studies: Critical Perspectives on Business and Management, Volume 1: 853–869. London and New York: Routledge.

Law, J. 1997a. Traduction/Trahison: Notes on ANT. Department of Sociology and Social Anthropology, Keele University, Keele, England. http://www.keele.ac.uk/depts/so/staff/jl/pubs-JL2.htm.

Law, J. 1997b. Topology and the Naming of Complexity. Department of Sociology and Social Anthropology, Keele University, Keele, England. http://www.keele.ac.uk/depts/so/staff/jl/pubs-JL3.htm.

Lewis, W. A. 1978. The Evolution of the International Economic Order. Princeton, NJ: Princeton University Press.

Lewis, W. A. 1966. Development Planning: The Essentials of Economic Growth. London: Allen and Unwin.

Lewis, W. A. 1955. The Theory of Economic Growth. London: Allen and Unwin.

Lieberman, I. 1990. Industrial Restructuring: Policy and Practice. Washington, DC: The World Bank.

Likert, R. 1967. The Human Organization. New York: McGraw-Hill.

Lim, L. 1985. Women Workers in Multinational Enterprises in Developing Countries. Geneva: ILO.

Lim, L. 1983. Capitalism, Imperialism, and Patriarchy: The Dilemma of Third World Women Workers in Multinational Factories. In J. Nash & M. P. Fernandez-Kelly (Eds.) Women, Men, and the International Division of Labor: 70–91. Albany, NY: SUNY Press.

Lim, L. 1981. Women's Work in Multinational Electronics Factories. In R. Dauber & M. L. Cain (Eds.) Women and Technology Change in Developing Countries: 181–192. Boulder, CO: Westview Press.

Lo, T. W., Yau, O. H. & Li, Y. 1986. International Transfer of Service Technology: An Exploratory Study of the Case of Supermarkets in China. Management International Review, 26 (2): 71–76.

Machungwa, P. D. & Schmitt, N. 1983. Work Motivation in a Developing Country. Journal of Applied Psychology, 68 (1): 31–42.

Mani, L. 1989. Multiple Mediations: Feminist Scholarship in the Age of Multinational Reception. Inscriptions, 5: 1–24.

Manzo. K. 1991. Modernist Discourse and the Crisis of Development Theory. Studies in Comparative International Development, 26 (2): 3–36.

March, J. & Simon, H. 1958. Organizations. New York: John Wiley and Sons.

Marcus, G. & Fischer, M. 1986. Anthropology as Cultural Critique. Chicago: Chicago University Press.

Martin, B. 1988. Feminism, Criticism, and Foucault. In I. Diamond and L. Quinby (Eds.) Feminism and Foucault: Reflections on Resistance. Boston: Northeastern University Press.

Martin, J. 1990. Deconstructing Organizational Taboos: The Suppression of Organizational Conflicts in Organizations. Organization Science, 1: 339–359.

Martinez, A. 1984. Improving Public Sector Management in Developing Countries. Government Accountants Journal, 33 (2): 37–40.

McAllister, C. 1991. Uneven and Combined Development: Dynamics of Change and Women's Everyday Forms of Resistance in Negeri Sembilan, Malaysia. Review of Radical Political Economics, 23 (3 & 4): 57–98.

McBeath, G. 1990. Practical Management Development. Cambridge, MA: Basil Blackwell.

McClintock, A., Mufti, A. & Shohat, E., (Eds.) 1997. Dangerous Liaisons: Gender, Nation, and Postcolonial Perspectives. Minneapolis: University of Minnesota Press.

McGregor, D. 1967. The Professional Manager. New York: McGraw-Hill.

McGregor, D. 1960. The Human Side of Enterprise. New York: McGraw-Hill.

McKendrick, N. 1962. Josiah Wedgewood and the Factory Discipline. The Historical Journal, 4: 30–35.

Meier, G. M. & Seers, D., (Eds.) 1984. Pioneers in Development. A World Bank Publication. New York: Oxford University Press.

Memmi, A. 1967. The Colonizer and the Colonized. Boston: Beacon Press.

Mintzberg, H. 1971. Managerial Work: Analysis from Observations. Management Science, October: 31–48.

Mintzberg, H. 1970. Structured Observation as a Method to Study Managerial Work. The Journal of Management Studies. February: 87–104.

Mir, R. A., Calás, M. B & Smircich, L. 1999. Global Technoscapes and Silent Voices: Challenges to Theorizing Global Cooperation. In D. Cooperrider & J. Dutton (Eds.) Organizational Dimensions of Global Change: 270–290. Thousand Oaks, CA: Sage.

Mitchell, T. 1995. The Object of Development. America's Egypt. In J. Crush (Ed.) Power of Development: 129–157, London: Routledge.

Mitchell, T. 1988. Colonising Egypt. Cambridge: Cambridge University Press.

Mohamad, M. 1997. Speech made at the Private Sector Salute to ASEAN in Petaling Jaya, Selangor Darul Ehsan, December 13th. http://www.smpke.jpm.my/pm/speeches.html.

Mohamad, M. 1993a. The Second Outline Perspective Plan, 1991–2000. In A. Sarji Abdul Hamid (Ed.) Malaysia's Vision 2020: Understanding the Concept, Implications and Challenges: 421–447. Selangor, Malaysia: Pelanduk Publications.

Mohamad, M. 1993b. Malaysia: The Way Forward. In A. Sarji Abdul Hamid (Ed.) Malaysia's Vision 2020: Understanding the Concept, Implications and Challenges: 403–420. Selangor, Malaysia: Pelanduk Publications.

Mohamad, M. 1986. Speech at the International Seminar on Commodities in Kuala Lumpur, Malaysia, July 21st. As published in Foreign Affairs Malaysia, 19 (3): 6–11.

Mohamad, M. 1982. The Malay Dilemma. Selangor, Malaysia: Federal Publications.

Mohanty, C. T. 1991. Cartographies of Struggle: Third World Women and the Politics of Feminism. In Mohanty, C. T., Russo, A. & L. Toures (Eds.) Third World Women and the Politics of Feminism: 1–47. Bloomington, IN: Indiana University Press.

Mohd Sheriff Mohd Kassim. 1993. Vision 2020's Linkages with the Sixth Malaysia Plan and the Second Outline Perspective Plan. In A. Sarji Abdul Hamid (Ed.) Malaysia's Vision 2020: Understanding the Concept, Implications and Challenges: 67–86. Selangor, Malaysia: Pelanduk Publications.

Molz, R. 1990. Privatization in Developing Countries. Columbia Journal of World Business, 25 (Spring/Summer): 17–24.

Morgan, G. 1986. Images of Organization. Newbury Park, CA: Sage.

Mosley, P., Harrigan, J. & Toye, J., (Eds.) 1991. Aid and Power: The World Bank and Policy-based Lending. Volume 2. London and New York: Routledge.

Mueller, A. 1987a. Peasants and Professionals: The Social Organization of Women in Development Knowledge. Unpublished Ph.D. dissertation. Department of Education, Ontario Institute for Studies in Education, University of Toronto.

Mueller, A. 1987b. Power and Naming in the Development Institution: The "Discovery" of "Women in Peru." Paper presented at the Fourteenth Annual Third World Conference, Chicago.

Nandy, A. 1989. Shamans, Savages and the Wilderness: On the Audibility of Dissent and the Future of Civilizations. Alternatives, 14 (3): 263–277.

Nandy, A. (Ed.) 1988. Science, Hegemony and Violence: A Requiem for Modernity. New York: Oxford University Press.

Nandy, A. 1983. The Intimate Enemy: Loss and Recovery of Self Under Colonialism. New York: Oxford University Press.

Nasif, E. G., Al-Daeaj, H., Ebrahimi, B. & Thibodeaux, M. S. 1991. Methodological Problems in Cross-Cultural Research: An Updated Review. Management International Review, 31 (1): 79–91.

National Productivity Corporation Malaysia (NPC). 1995. P and Q Enhancement Programs. Petaling Jaya, Selangor.

National Productivity Corporation Malaysia (NPC). 1994. Malaysia Productivity Report. Petaling Jaya, Selangor.

National Productivity Corporation Malaysia (NPC). 1992. Annual Report. Petaling Jaya, Selangor.

National Productivity Corporation Malaysia (NPC) & Standards and Industrial Research Institute of Malaysia (SIRIM). 1995. The 7 New QC Tools. [Training flyer], Petaling Jaya, Selangor.

Negandhi, A. R. 1985. Management in the Third World. In P. Joynt & M. Warner (Eds.) Managing in Different Cultures: 69–97. Oslo, Norway: Universitetforlaget.

Nkomo, S. M. 1992. The Emperor has no Clothes: Rewriting "Race" in Organizations. Academy of Management Review, 17: 487–513.

Nooteboom, B., Berger, H. & Noorderhaven, N. G. 1997. Effects of Trust and Governance on Relational Risk. Academy of Management Journal, 40 (2): 308–338.

Nowotny, H. 1976. Time Structuring and Time Measurement. In J. T. Fraser, N. Lawrence & D. Park (Eds.) The Study of Time. Volume 2. New York: Springer.

Ohmae, K. 1993. The Rise of the Region State. Foreign Affairs, 71 (2): 78–87.

Ohmae, K. 1990. The Borderless World. London: Collins.

Ohmae, K. 1989a. The Global Logic of Strategic Alliances. Harvard Business Review. 67 (2): 143–154.

Ohmae, K. 1989b. Managing in a Borderless World. Harvard Business Review, May/June.

Ong, A. 1993. The Gender and Labor Politics of Postmodernity. Annual Review of Anthropology, 20: 279–309.

Ong, A. 1987. Spirits of Resistance and Capitalist Discipline: Factory Women in Malaysia. Albany, NY: SUNY Press.

O'Reilly, A. J. F. 1988. Establishing Successful Joint Ventures in Developing Nations: A CEO's Perspective. Columbia Journal of World Business, 23 (1): 65–71.

Osborn, R. N. & Hagedoorn, J. 1997. The Institutionalization and Evolutionary Dynamics of Interorganizational Alliances and Networks. Academy of Management Journal, 40 (2): 261–278.

Palmer, M., Ramsey, J. & Kwok, M. H. 1995. Kuan Yin. London: Thorsons.

Pandey, G. 1990. The Construction of Communalism in Colonial North India. New Delhi, India: Oxford University Press.

Parajuli, P. 1991. Power and Knowledge in Development Discourse. International Social Science Journal, 127: 173–190.

Park, S. H. & Ungson, G. R. 1997. The Effect of National Culture, Organizational Complementarity, and Economic Motivation on Joint Venture Dissolution. Academy of Management Journal, 40 (2): 279–307.

Parkhe, A. 1993. Strategic Alliance Structuring: A Game Theoretic and Transaction Cost Examination of Interfirm Cooperation. Academy of Management Journal, 36 (4): 794–829.

Parkhe, A. 1991. Interfirm Diversity, Organizational Learning, and Longevity in Global Strategic Alliances. Journal of International Business Studies, 22: 579–601.

Paul, S. 1984. The Strategic Management of Development Programs: Evidence from an International Study. International Studies of Management and Organization, 14 (2–3): 113–166.

Paul, S. 1982. Managing Development Programs: Lessons for Success. Boulder, CO: Westview Press.

Pfeffermann, G. P. & Madarassy, A. 1992. Trends in Private Investment in Developing Countries 1993: Statistics for 1970–1991. Washington, DC: The World Bank.

Porter, M. 1990. The Competitive Advantage of Nations. Harvard Business Review, March/April.

Porter, M. 1986. Changing Patterns of International Competition. California Management Review, 28 (2): 9–40.

Porter, M. 1985. Competitive Advantage. New York: Free Press.

Prakash, G. 1990. Writing Post-Orientalist Histories of the Third World: Perspectives from Indian Historiography. Society for Comparative Study of Society and History. No. 0010–4175/90/2893.

Psacharopoulos, G. 1984. Assessing Training Priorities in Developing Countries: Current Practice and Possible Alternatives. International Labor Review, 123 (5): 569–583.

Putnam, L. 1996. Situating the Author and Text. Journal of Management Inquiry, 5: 382–386.

Putti, J. M., Aryee, S. & Tan, T. K. 1989. Work Values and Organizational Commitment: A Study in the Asian Context. Human Relations, 42 (3): 275–288.

Rabinow, P. 1986. Representations are Social Facts: Modernity and Post-Modernity in Anthropology. In J. Clifford and G. Marcus (Eds.) Writing Culture: The Poetics and Politics of Ethnography: 234–261. Berkeley: University of California Press.

Radhakrishnan, R. 1996. Diasporic Mediations: Between Home and Location. Minneapolis: University of Minnesota Press.

Radhakrishnan, R. 1994. Postmodernism and the Rest of the World. Organization, 1 (2): 305–340.

Radhakrishnan, R. 1993. Postcoloniality and the Boundaries of Identity. Callaloo, 16 (4): 750–771.

Rahnema, M. 1988a. Power and Regenerative Processes in Micro-Spaces. International Social Science Journal, 117: 361–376.

Rahnema, M. 1988b. A New Variety of AIDS and its Pathogens: Homo Economicus, Development and Aid. Alternatives, 13 (1): 117–136.

Rashid, R. 1993. A Malaysian Journey. Petaling Jaya, Malaysia: Rehman Rashid.

Ratnam, K. J. 1965. Communalism and the Political Process in Malaya. Singapore: University of Malaya Press.

Reed, M. 1999. From the "Cage" to the 'Gaze'? The Dynamics of Organizational Control in Late Modernity. In G. Morgan & L. Engwall (Eds.) Regulation and Organizations: International Perspectives: 17–49. London: Routledge.

Rich, A. 1984. Notes Towards a Politics of Location. Lecture for the first session of the Utrecht Summer School of Critical Semiotics, June 1, University of Utrecht. In M. Díaz-Diocaretz and I. M. Zavala (Eds.) Women, Feminist Identity and Society in the 1990s: Selected Papers. Amsterdam: John Benjamins, 1985.

Ricks, D. A., Toyne, B. & Martinez, Z. 1990. Recent Developments in International Management Research. Journal of Management, 16 (2): 219–253.

Rieger, F. & Wong-Rieger, D. 1988. Model Building in Organizational/Cross-cultural Research: The Need for Multiple Methods, Indices, and Cultures. International Studies of Management and Organization, 18 (3): 19–30.

Roberts, K. & Boyacigiller, N. A. 1984. Cross-National Organizational Research: The Grasp of the Blind Men. Research in Organization Behavior, 6: 423–475.

Rofel, L. 1992. Rethinking Modernity: Space and Factory Discipline in China. Cultural Anthropology, 7 (1): 93–114.

Ronen, S. 1986. Comparative and Multinational Management. New York: John Wiley.

Ronen, S., & Shenkar, O. 1985. Clustering Countries on Attitudinal Dimensions: A Review and Synthesis. Academy of Management Review, 10 (3): 435–454.

Rostow, W. W. 1960. Stages of Economic Growth. Cambridge: Cambridge University Press.

Roy, D. F. 1960. Banana Time: Job Satisfaction and Informal Interaction. Human Organization, 18: 156–168.

Rushdie, S. 1992. Imaginary Homelands. New York: Penguin.

Sachs, W. (Ed.) 1992. The Development Dictionary. A Guide to Knowledge as Power. London: Zed Books.

Sachs, W. 1992. Introduction. In W. Sachs (Ed.) The Development Dictionary. A Guide to Knowledge as Power: 1–6. London: Zed Books.

Safavi, F. 1981. A Model of Management Education in Africa. Academy of Management Review, 6(2): 319–331.

Said, E. 1989. Representing the Colonized: Anthropology's Interlocutors. Critical Inquiry, 15: 205–225.

Said, E. 1983. The World, the Text, the Critic. Cambridge, MA: Harvard University Press.

Said, E. 1979. Orientalism. New York: Vintage Books.
Sardar, Z. (Ed.) 1988. The Revenge of Athena: Science, Exploitation and the Third World. London: Mansell Publishing Limited.
Sarji Abdul Hamid, A. (Ed.) 1995. Malaysia's Vision 2020: Understanding the Concept, Implications and Challenges. Selangor, Malaysia: Pelanduk Publications.
Sarji Abdul Hamid, A. 1994. The Civil Service of Malaysia: A Paradigm Shift. A Government of Malaysia publication. Kuala Lumpur, Malaysia: Percetakan Nasional Malaysia Berhad.
Sarji Abdul Hamid, A. 1993. Improvements and Development in the Public Service for the Year 1992. Kuala Lumpur, Malaysia: Percetakan Nasional Malaysia Berhad.
Schneider, S. 1989. Strategy Formulation: The Impact of National Culture. Organization Studies, 10 (2): 149–168.
Schwartz, H. 1988. The Potential Role of Behavioral Analysis in the Promotion of Private Enterprises in Developing Countries. Columbia Journal of World Business, 23 (1): 53–56.
Scott, J. C. 1985. Weapons of the Weak: Everyday Forms of Peasant Resistance. New Haven, CT: Yale University Press.
Segura, E. 1988. Industrial, Trade and Financial Sector Policies to Foster Private Enterprises in Developing Countries. Columbia Journal of World Business, 23 (1): 19–26.
Sekaran, U. 1983. Methodological and Theoretical Issues and Advances in Cross-Cultural Research. Journal of International Business Studies, 14 (2): 61–73.
Sheth, D. L. 1987. Alternative Development as Political Practice. Alternatives, 12 (2): 155–171.
Shiva, V. 1990. The Violence of the Green Revolution. Third World Agriculture, Ecology and Politics. London: Zed Books.
Shiva, V. 1989. Staying Alive: Women, Ecology and Development. London: Zed Books.
Shiva, V. 1987. The Violence of Reductionist Science. Alternatives, 12 (2): 243–261.
Silverman, D. 1993. Interpreting Qualitative Data: Methods for Analysing Talk, Text, and Interaction. London: Sage Publications.
Smircich, L., Calás, M. B. & Morgan, G. (Eds.) 1992. New Intellectual Currents in Organization and Management Theory. Academy of Management Review, 17 (3): 404–406.
Smith, D. E. 1990. The Conceptual Practices of Power. Boston: Northeastern University Press.
Smith, D. E. 1987. The Everyday World as Problematic. Boston: Northeastern University Press.
Sood, A. & Kohli, H. 1985. Industrial Restructuring in Developing Countries. Finance and Development: 46–49.
Spivak, G. 1990. The Postcolonial Critic: Interviews, Strategies, Dialogues. S. Harasym (Ed.). London and New York: Routledge.
Spivak, G. 1987. Subaltern Studies: Deconstructing Historiography. In In Other Worlds: Essays in Cultural Politics: 197–221. New York: Methuen.

Standards and Industrial Research Institute of Malaysia (SIRIM). 1995. SIRIM Profile. Selangor, Malaysia.

Standards and Industrial Research Institute of Malaysia (SIRIM). 1994. SIRIM Annual Report. Selangor, Malaysia.

Stephens, G. K. & Grier, C. R. 1995. Doing Business in Mexico: Understanding Cultural Differences. Organizational Dynamics, Summer: 39–55.

Stewart, S., Cheung, M. T. & Yeung, D. W. K. 1992. The Latest Asian Newly Industrialized Economy Emerges: The South China Economic Community. Columbia Journal of World Business, 27 (2): 30–37.

Tan, S. B. 1992. Counterpoints in the Performing Arts of Malaysia. In J. S. Kahn & F. Loh (Eds.) Fragmented Vision: Culture and Politics in Contemporary Malaysia: 282–305. Sydney: Allen and Unwin.

Taussig, M. 1980. The Devil and Commodity Fetishism in South America. Chapel Hill, NC: University of North Carolina Press.

Tayeb, M. 1987. Contingency Theory and Culture: A Study of Matched English and Indian Manufacturing Firms. Organization Studies, 8 (3): 241–261.

Taylor, F. W. 1911. The Principles of Scientific Management. New York: Harper.

Teece, B. 1989. Profiting from Technological Innovation: Implications for Integration, Collaboration, Licensing and Public Policy. Research Policy, 15: 285–305.

Thich Nhat Hanh. 1998. Fragrant Palm Leaves. Journals 1962–1966. Berkeley: Parallax Press.

Thompson, E. P. 1966. The Making of the English Working Class. New York: Vintage.

Thompson, J. D. 1967. Organizations in Action. New York: McGraw-Hill.

Thurow, L. 1992. Head to Head: The Coming Economic Battle among Japan, Europe, and America. New York: William Morrow and Company, Inc.

Torbion, I. 1985. The Structure of Managerial Roles in Cross-cultural Settings. International Studies of Management and Organization, 15 (1): 52–74.

Touraine, A. 1995. Critique of Modernity. Cambridge, MA: Blackwell.

Townley, B. 1994. Reframing Human Resource Management: Power, Ethics and the Subject at Work. London: Sage.

Townley, B. 1993. Foucault, Power/Knowledge, and its Relevance for Human Resource Management. Academy of Management Review, 18: 518–545.

Trinh T. Minh-ha 1989. Women, Native Other. Bloomington, IN: Indiana University Press.

Usinier, J. C. 1991. Business Time Perspectives and National Cultures: A Comparative Study. Management International Review, 31: 197–217.

Vernon-Wortzel, H. & Wortzel, L. H. 1988. Globalizing Strategies for Multinationals from Developing Countries. Columbia Journal of World Business, 23 (91): 27–34.

Visvanathan, S. 1987. From the Annals of the Laboratory State. Alternatives, 12 (1): 37–60.

Walters, R. S. & Blake, D. H. 1992. The Politics of Global Economic Relations. Fourth edition. Englewood Cliffs, NJ: Prentice Hall.

Warner, M. 1986. Managing Human Resources in China: An Empirical Study. Organization Studies. 7 (4): 353–366.

Weber, M. 1947. The Theory of Social and Economic Organizations. Translated by A. M. Henderson & T. Parsons. New York: Oxford University Press.

Weigel, D. R. 1988. Investment in LDCs: The Debate Continues. Columbia Journal of World Business, 23 (1): 5–9.

World Bank. 1993a. The East Asian Miracle: Economic Growth and Public Policy. A World Bank Policy Research Report. Washington, DC.

World Bank. 1993b. Getting Results: The World Bank's Agenda for Improving Development Effectiveness. Washington, DC.

World Bank. 1991. Developing the Private Sector: The World Bank's Experience and Approach. Washington, DC.

World Bank. 1991. World Development Report. Washington, DC.

World Bank. 1989. Malaysia: Matching Risks and Rewards in a Mixed Economy Program. A World Bank Country Study. Washington, DC.

Yavas, U. & Cavusgil, T. 1989. Management know how Transfer to Developing Countries: Efficacy of Specialized Training Courses. Management International Review, 29 (3): 72–80.

Yavas, U., Kaynak, E. & Dilber, M. 1985. The Managerial Climate in Less-Developed Countries. Management Development, 23 (3): 29–40.

Yip, G. 1995. Global Strategy . . . In a World of Nations? In C. A. Bartlett & S. Ghoshal (Eds.) Transnational Management. Boston: Irwin McGraw-Hill.

Index

E

F

G

H

P

Q

R

S

For Product Safety Concerns and Information please contact our EU representative GPSR@taylorandfrancis.com
Taylor & Francis Verlag GmbH, Kaufingerstraße 24, 80331 München, Germany

www.ingramcontent.com/pod-product-compliance
Ingram Content Group UK Ltd.
Pitfield, Milton Keynes, MK11 3LW, UK
UKHW022001180926
475701UK00003B/139

* 9 7 8 0 4 1 5 6 4 9 8 5 8 *